HARVARD EAST ASIAN MONOGRAPHS

53

UNDERSTANDING BUSINESS CONTRACTS IN CHINA

1949–1963

UNDERSTANDING BUSINESS CONTRACTS IN CHINA, 1949–1963

by

Richard M. Pfeffer

Published by

East Asian Research Center

Harvard University

Distributed by

Harvard University Press

Cambridge, Mass.

1973

Copyright, 1973, by

The President and Fellows of

Harvard College

The East Asian Research Center at Harvard University
administers research projects designed to further
scholarly understanding of China, Japan, Korea,
Vietnam, and adjacent areas. These studies have been
assisted by grants from the Ford Foundation.

Library of Congress No. 70-162859
SBN 674-92095-3

To my parents, Harry and Syril Pfeffer

CONTENTS

PREFACE

America and I became politically self-conscious during the 1960s. When I began in the early sixties to study and write about contracts in China, I was a believer. I believed, with only the usual liberal reservations, in the mythologies of the American political and legal systems. And relatedly, I tended to see the Chinese system from the liberal perspective, decrying government control, manipulation, and the lack of liberal tolerance and freedoms in China. My beliefs, of course, were reflected, subtly or not so subtly, in my scholarship: in the topics chosen; in the questions asked, and those not asked; and in the language and methodologies used.

Because this book has been composed and is being published as a series of essays, researched and written over a period of eight years (1961–1969), it reflects, more than most books perhaps, its author's development—where he began and how he changed. Consequently, this volume incorporates two quite different approaches to understanding contracts in China. One, manifest in chapters II, III, and IV, came first chronologically and involves conventional comparative analysis of Chinese contracts. The other, seen in chapters I, V, and VI, returns to the source of my comparative perspectives and re-examines the Anglo-American theory and tradition of contracts. This re-examination suggests that conventional comparative analysis has been biased and misconceived.

Chapters II, III, and IV, written several years before the other chapters and published here substantially as they appeared previously in article form, are products of early presuppositions. Chapters II and III were written in 1962 and were published in numbers 14 and 15 of *The China Quarterly* (April-June 1963 and July-September 1963). Chapter IV was written next, in 1965, and was published in number 28 of *The China Quarterly* (October-December 1966).

As the introduction will suggest, my work on contracts in China began with a series of presuppositions derived generally from my liberal background and specifically from my training as a lawyer at the Harvard Law School. I began legalistically and ethnocentrically, presupposing, for example, both the reality and universal validity of the free market economy in the United States and the relevance of the stereotypical role that contracts were said to play in that economy. By 1964, I was shifting disciplines from law to political science and was becoming somewhat more intellectually self-conscious. By 1967, when I returned to work again on contracts after a hiatus of two years, I had moved leftward politically. My skepticism was becoming more diffuse, spreading from a mere lack of faith in particular men, in particular American institutions, and in particular schools of liberal scholarship. That new skepticism is reflected to a degree in sections of the penultimate and concluding chapters of this volume, which argue that conventional wisdom has failed to understand the role of contracts in societal development and the present role of contracts in the United States and in China. The manuscript was completed in 1969, after Chicago '68.

This book, then, reflects an incomplete journey from one set of beliefs and assumptions through skepticism toward another set. As I was not yet in 1969 a new believer, the concluding chapter is composed largely of questions and speculation rather than definitive statements.

The book itself, moreover, is not definitive. It is, frankly, a first step, based upon a finite amount of evidence reasonably collected and sifted, and upon considerable thought.

It is not definitive of the Chinese side for the 1949–1963 period covered, since there have been few studies of Chinese contracts upon which such a work could be built. Even studies of related areas, like the day-to-day functioning of the decentralized planning institutions in China, remain quite limited.

Nor is the book conclusive on the American side, for it is based on the small number of secondary sources that focus on relevant socioeconomic developments in contract practice in America. In light of the dearth of secondary literature on the practice of contracts in America, I have had to rely heavily on two sources, Lawrence Friedman's excellent and controversial *Contract Law in America* and Stewart Macaulay's provocative "Non-contractual Relations in Business: A Preliminary Study." Both are studies of contract practices in Wisconsin. Consequently the arguments in chapters I and V about contracts in America that are derived from these sources should not be taken as an empirical refutation of the still dominant stereotypes of nationwide contract practice.

Moreover, even within the bounds of such modesty, this book does not purport to be comprehensive concerning business contract practice in America. Rather, it is selective. It does not, for example, analyze the way business contracts increasingly are handled as part of state commercial law by the Uniform Commercial Code. Nor does it delve into recent court decisions relating to contracts, except insofar as such decisions have been incorporated in the sociologically-oriented, selected secondary literature.

There is, on the other hand, one seemingly trivial way in which this volume is authoritative. It reflects its author's intellectual development. That in itself may be significant if, as I believe, my development during the 1960s was typical of certain broader, contemporaneous developments in the United States within the China field and within the legal profession.

In both fields in the sixties a "new breed" began to challenge the preconceptions and practices of those who had shaped the fields' assumptions. Preconceptions about our system, which frequently had done double duty, both as legitimating myths for policymakers and as self-satisfying frameworks for comparative analysis of other societies' institutions, have increasingly been taken for just what they are, myths. The United States, for example, no longer—if it ever did—looks more rational or less ethnocentric than China. Questions about the quality of justice in America no longer can be turned aside by facile references to "totalitarianism."

Those of us, whatever our age, who grew up intellectually and morally in the 1960s were dis-enchanted. Many of us began to suspect that the presuppositions handed down as conventional wisdom in the China field serve an imperialist American foreign policy and that that service is more than coincidental, if less than conspiratorial. We came to realize that America's stated concern for due process is inadequately reflected in its own criminal enforcement practice and that due process standards, in any event, are inappropriate for judging the legal systems of societies in revolution. Many of us, in short, discovered American ethnocentrism in the service of American power, rationalized by American liberal scholarship. And in response we tried to ask different questions and to seek alternative frameworks for analysis.

Few of us, thus far, have found final, or even satisfying answers in this quest. We have, however, been confirmed in our feeling that without the quest we would have remained even more complacent, parochial, and self-righteous in our scholarship and in our lives than we may now be.

This book was begun, then, in 1961, when I was a liberal believer, and was completed in 1969, in a state of cynicism about the American reality. Had I in 1969, based upon my new and very tentative conclusions about the practice of contracts in America, returned to review the Chinese case, I certainly would have asked different questions than I did in 1962 or in 1964. As a consequence, I undoubtedly would have used different frameworks for analysis and arrived at somewhat different conclusions in Chapters II and III. Even Chapter IV might well have read quite differently.

But whatever substantial gains might have accrued from such a re-investigation of Chinese contracts, the process of one man's intellectual development and the significance of such development to the form and substance of his work would have been obscured. Consistent theses about contracts would have been presented as "the truth." Instead, a number of theses, which may be inconsistent, are offered here. It would be neater were it not so, but it would not be as honest, nor necessarily a more useful portrayal of a complex institution.

Be that as it may, I feel I have served my time on contracts, and have passed on to something else. It will be up to someone else to produce a work more definitive. I hope that this book will be of some use in such an endeavor.

Whatever the book's merits and defects—for both of which I take ultimate responsibility—I am grateful to many friends and colleagues, as well as to several foundations and agencies, for the intellectual, psychological, financial, and mechanical help they have given me.

My primary professional debt is to Jerome A. Cohen, Professor at the Harvard Law School, who has aided me in every way imaginable. I owe a similar debt to John K. Fairbank, who has always provoked me. I am also grateful to a number of other professors, including Barrington Moore, Jr., Ezra Vogel, Benjamin I. Schwartz, Arthur von Mehren, Dwight Perkins, Harold J. Berman, and Samuel P. Huntington, all of Harvard, and to Howard Zinn of Boston University.

Without grants from the Ford Foundation, the federal government (NDFL), the East Asian Research Center at Harvard, and Harvard Law School, and the Law School at the University of Chicago, I could not have undertaken nor completed this work. Particular appreciation is due Phil Neal, Dean of the University of Chicago Law School, who allowed me complete freedom to do my own work during my half-time fellowship there in 1967–1969.

I also want to express the deepest debts of gratitude to several close friends and relatives who helped in many ways to sustain me in my work and life during the completion of this manuscript. I owe much to my parents, Harry and Syril Pfeffer, to Richard Rubenstein (Roosevelt University), to Libby Rubenstein, to Margo Pfeffer, to Allan Chasanoff, to Marvin Zonis (University of Chicago), to Eqbal Ahmad (Institute for Policy Studies), to my wife, Elinor Bacon, and to Soren, my son. Finally, I owe an intellectual and moral debt to my comrades in the Committee of Concerned Asian Scholars who have contributed to my awakening and to the organization itself, which makes me proud to be in the China field.

There are many others from whom I have learned much, too frequently not enough. But the list already is so long as to overdignify the product, if not the feeling.

Richard M. Pfeffer

Johns Hopkins University
January 3, 1973

INTRODUCTION

Institutions are infinitely complex, messy, and internally contradictory—a painfully obvious observation, but one inherently antithetical to the tendency of social sciences to explain reality "systematically."

The questions with which one comes to analysis in large part shape the conclusions of the analysis, however rigorous one may try to be in keeping the posing of initial hypotheses tentative and the collection of data appropriately open-ended. Another truism, but one peculiarly relevant to the analysis of institutions in societies we have been educated to see as our "enemy."[1]

Comparative analysis of an alien society's law-related institutions is especially prickly, for the social mythologies about our own legal system that too frequently limit the perspectives and values involved in such analysis have been perpetuated and reinforced by our courts, our specialized (dis-integrated) legal treatises, and our legal education. An American-trained legal scholar is likely to have an aggrandized vision of the role and the quality of formal legal institutions in American society. And it is with that vision, amounting almost to a trained incapacity, that he begins comparative analysis.[2]

Comparative analysis of business contracts is, perhaps, most difficult of all, for the American analyst must deal with two additional major, if not insurmountable, problems. First, Anglo-American contract theory is a central part of American liberal ideology. While it is clear that the whole regime of property and its transactional counterpart, contracts, constitutes a "system of legitimized coercion" in which "the line must be drawn, not between pressure or no pressure, but between permissible and non-permissible pressure,"[3] contract is generally associated in our tradition with freedom. For the most part, our law of contracts has been developed and abstractly systematized as if the American economy were a free market economy, pure and simple, and as if, in any event, that state were devoutly to be desired. Second, almost no empirical work has been carried out regarding the actual role of formal business contracts in America. Thus, while there have been intimations that theory and practice might diverge greatly, contract theory, rather than practice, has set the perspectives for comparative analysis.

All these problems exist and are compounded by others when the object for comparison is contracts in Communist China. For, if we know little of American business practice (outside the practice reflected in cases that have been adjudicated), we know even less of the Chinese practice, and very little even about the Chinese theory of contracts. Moreover, what we do know of contract theory in China tends to come from a period when the Russian model of economic development was more comprehensively dominant and when Soviet legal thought and institutions tended to be taken as models for

China. Finally, the velocity of change in Chinese institutions appears to be relatively high. So, the significance of contracts in China cannot be tied, except as a matter of history, to particular institutions.

And yet there is reason to think that the study of contracts in China is an important task, but perhaps not for reasons one might suppose, given the Anglo-American tradition. The Chinese themselves continue to appear to believe that contracts are a valuable institution. We must try to understand why.

Moreover, the study of contracts in a setting so alien to us may suggest that our legalistic vision of contracts is quite narrow and culture-bound. It may help to highlight the fact that full-blown freedom of contract was a relatively recent development in Anglo-American history and was dominant only for a relatively short period. It may suggest that other societies in their development process need not repeat our stages of development, nor be limited to utilizing contracts in the American way.

At some point, the study of contracts in China may raise the issue whether, spread so thin, "contract" today has any effective core meaning. But it is appropriate that that question be raised, as it is being raised increasingly in practice by developments in America itself during the last fifty years.

All these issues and others will be dealt with in this book, but to some extent they will not be fitted within a single consistently applied perspective. Each chapter contains perspectives and arguments that tell part of the truth and inevitably distort the rest. In a sense, this book includes an implied thesis-within-a-thesis (more successful, I hope, because more modest, than the revolution within the revolution). The "thesis within" is that contracts are not what we have thought or taught. This thesis involves a demonstration *by example* of some problems of comparative analysis, as reflected in the proposition that what one seeks determines what one finds. Rather than homogenizing these differences into a single perspective that might boil down to "on-the-one-hand-but-on-the-other," I have chosen to leave the differences less resolved. In addition to the "thesis within," several unifying themes, such as the relationship between freedom and authority in contracts and the role of contracts and formal legal institutions in various economies and in development, will be explored.

The first chapter discusses broadly the nature of contracts, their Anglo-American development, and their relation to economic development in the West. To some extent it prefigures Chapter V in going beyond more classical Anglo-American conceptions of contracts and contract law formulated in the eighteenth and nineteenth centuries.

The second, third, and fourth chapters deal with China and are written generally from the perspective of classical Anglo-American contract theory as enunciated by our courts. Within that perspective, it is natural that the first question to be asked in analyzing Chinese contracts concerns the role of contracts in a planned economy, for tradition has it that contracts in Anglo-American history essentially express the will of individual parties rather than broader societal concerns. The second and fourth chapters primarily deal with contract in presumed confrontation with legal duties imposed by plans.

The third chapter, remaining also within the classical tradition, analyzes, in terms of their respective functional relations to economies taken to be fundamentally different, the Anglo-American theory of contracts and the Chinese theory of contracts. In both the second and third chapters, true to the Anglo-American tradition, it is assumed that the legal implications of contracts are dominant and that the legal rules, therefore, are very significant, or at least in the case of China, likely to be so, in the near future. This argument, like legal science in general, is convincing in its own terms, but probably a half-truth at best. These chapters begin to suggest that in China there is no single institution of business contracts, but different sorts of contracts—a conclusion that appears to be equally applicable today to the American scene but one that is beyond the understanding of traditional contract theory.

The fourth chapter tests the significance of Chinese contract theory in practice through an institutional case study of business contracts in agriculture. It finds the significance of formal rules and formal methods of dispute resolution to be minimal.

The fifth and final chapters bring the book full circle, having been provoked, ironically, by the findings regarding China to question the meaning and significance of contracts in America and China today and, more broadly, to speculate on the role of business contracts in economic development.

Chapter I

CONTRACTS IN THE ANGLO-AMERICAN TRADITION

The institution of contracts, as we have come to understand it in the West, is highly complex, varied in its manifestations, and exceedingly difficult to define.[1] Contracts involve elements in tension, such as freedom and authority, the varied combinations of which simultaneously give the institution its unique characteristics and its elusive quality. For purposes of analysis it is useful to postulate a core meaning for contracts, to definitionally maximize the differentia between it and other legal institutions. In formulating this ideal type of contracts as a formal, legal institution, the similarities and interconnections between contracts and other societal institutions may be obscured, but I shall return to this critical issue in Chapter V. For present purposes, it will suffice to establish the distinctive nature of contracts in the Anglo-American legal system.

The beginning, if not the end, of understanding the institution of contracts lies in its historical roots, its development, and its relation to the economic system in the West, of which it has been an integral part. The truism that law and legal institutions are products of their social and economic environment and that they change in response to environmental pressures is, perhaps, nowhere more clearly confirmed than in the case of contracts.

Historically, new institutions probably come to be understood, initially at least, in terms of how they diverge from previously existing institutional patterns. Thus, it appears that contracts came to be understood primarily as a divergence from traditional and feudal status-oriented institutions.[2] The focus in contract has been on the legal relationship between persons achieved by the voluntary agreement of the immediate parties, as reflected, to the extent possible, exclusively in the particular agreement itself. In emphasizing the *voluntary* nature of such agreements, the distinction between contracts and inherited, *in*voluntary personal status is brought out. And in emphasizing the *particular* relation based on the agreement at issue, the distinction is made between contracts and more *diffuse* status relations.

In this perspective what seems to be distinctive about contracts is precisely what twentieth-century sociologists might characterize as their "universalism" and "functional specificity." These qualities, and their relationship to a free market economy, are brilliantly described in a recent empirical case study of the institution of contracts in Wisconsin, from which I quote extensively:

Basically. . .the "pure" law of contract is an area of what we call abstract relationships. "Pure" contract doctrine is blind to details of subject matter and person. It does not ask who buys and who sells, and what is bought and sold. In the law of contract it does not matter whether the subject of the contract is a goat, . . .a carload of lumber, a stock certificate, or a shoe. As soon as it matters—*e.g.*, if the sale is of heroin,. . .or labor for twenty-five cents an hour—we are in one sense no longer talking pure contract. In the law of contract, it does not matter if either party is a woman, a man, . . .a corporation, the government, or a church. Again, as soon as it does matter. . .we are no longer talking pure contract. When the relationship of parties to land is treated as creating distinctive legal issues, simply because land is involved, this is. . .property law,. . .not contract. In contract cases land is treated as a commodity on the market, the same as every other commodity, and the rules are supposed to be the same as the rules for horses and cows. *Contract law is an abstraction*—what is left in the law relating to agreements when all particularities of person and subject matter are removed.

The abstraction of classical contract law is not unrealistic; it is a deliberate relinquishment of the temptation to restrict untrammeled individual autonomy or the completely free market in the name of social policy. *The law of contract is, therefore, roughly coextensive with the free market.* Liberal Nineteenth Century economics fits in neatly with the law of contracts so viewed. It, too, had the abstracting habit. In both theoretical models—that of the law of contracts and that of liberal economics—parties could be treated as individual economic units which, in theory, enjoyed complete mobility and freedom of decision. All goods could be treated as substitutable, one for the other, in money terms. All claims. . .were assignable. All rights could be valued in monetary. . .terms.

In fact, there was never any point at which the law of contract corresponded exactly with such an economic theory. But in a rough way, the rise and fall of the law of contract paralleled the rise and fall of liberal economics as a working philosophy. Mention of a few of the doctrines of the law of contract shows how closely the two ideological systems fit together. The classical [statement of] the doctrine of consideration. . .rejected the notion that any price fixed freely by two parties might be condemned in law as unfair (inadequate); only the market, as evidenced by what a willing seller paid a willing buyer, measured value. The law of contract damages took into account only generalized types of economic damage, ignoring any personal element (for example, embarrassment or humiliation resulting from breach of contract). Recoverable damages for breach of contract to deliver goods were computed on the basis of the difference between contract price and market price at the moment of breach. This formula assumed a frictionless and perfect market, operating instantaneously and universally. Conversely, the law of contract would not enforce contracts for which no market price was conceivable (for example, a contract so vague or one-sided that it had no ascertainable value) or purely personal agreements (such as a promise to attend a social function). The correspondence between law and economic theory was never exact. Contract law was not a book written by Adam Smith. Nobody purposely sat down to turn contract law into an applied branch of liberal economics. But a free market developed, and grew; the law of contract was the legal reflection of that market and naturally took on its characteristics. Contract was abstract; and the two institutions directed behavior along similar channels.[3]

This rough congruence between the "pure" law of contract and the free market—both being founded on abstraction—implies the shared boundaries of both: the boundaries are "where abstraction ends and particularity begins." This means that

> if we start (perhaps naively) with the idea that the law of contracts is coextensive with the universe of enforceable agreements, then any positive rule of law or statute which subtracts from or adds to the universe of allowable terms or agreements narrows or expands the potential area of the law of contracts. *The law of contract, then, is residuary*; it applies to those agreements that are not subject, in whole or in part, to special legal treatment, by virtue of some special statute or legal rule (the tax law, the statute on union membership clauses, the anti-trust law).[4]

An example of how this residuary may be expanded is provided by the decline in the significance of real property law (land law) and the correlative increase in the significance of contract law in the period from medieval times to the twentieth century.

The Heyday of Contracts

Despite the fact that most of us take contracts for granted as part of our daily life, the great significance attributed to contracts is a rather recent development. Blackstone, for example, whose classic on the English common law, *Commentaries,* was published in 1765 and who was a contemporary of Adam Smith, did not consider contract law important enough to give it much more than thirty pages in a four-book work, and, at that, Blackstone treated contracts as a "mere appendix to the 'Law of Things.'"[5]

Blackstone's perception of the intimate relationship between contracts and property recognized property rights as the *sine qua non* of contracts. Something that cannot be "owned" cannot be transferred by contract. But ownership of a property right, while necessary, is not a sufficient condition for the development of contracts. The rules of ownership, as was the case with medieval English land law, may allow little latitude for voluntary dispository agreements:

> The classic common law [of real property] recognized primogeniture: it gave the land to the descendant's oldest son, who inherited by virtue of his *status* as oldest son. Volition, intention, had nothing to do with the matter. Naturally, medieval land law was no chance development; a society organized in the manner of medieval England required a system of law which, compared to Nineteenth Century law, was more rigid, less concerned with volition, more formal, technical, procedural, status-conscious. . .Changes in the structure of land law favoring private contract were the sort of change which has been characterized. . .as the movement in law from status to contract.[6]

As land increasingly could be exchanged free of legal restraints, it became in the mid-nineteenth-century heyday of contracts one of the major commodities exchanged by contract. In nineteenth-century Wisconsin, for example, "land was treated as a commodity traded on the open market," thereby greatly expanding the subject matter and significance of contracts.[7] It was this sort of expansion of the subject matter that contracts might encompass more than any refinement in contract theory that led to the unprecedented prominence of contracts in the nineteenth century.

The systematization of the law of contracts in treatises and case books essentially reflected, rather than produced, these developments. It is ironic, in fact, that the nineteenth-century great works on contracts presented a recapitulation of the development of contract case law, which almost from the moment of publication increasingly came to be unrepresentative to the point of irrelevance of contracts in practice. The residuary, expanded during the earlier development of capitalism, increasingly was restricted during the development of the advanced industrial state.[8] And the restatement of the rules of contract law as if abstraction still reigned distorted reality.

Expediency and Ideology

Nevertheless, the field of contracts developed under the dominant liberal perspective that free men generally have the right to make whatever agreements they choose and that the legal system, for the most part, will enforce such agreements as made. This perspective had a utilitarian and an ideological basis.

The utilitarian aspect of freedom of contract was related to the rapid development of commerce and industry. Contract, as the product or counterpart of the free market, was useful to an economy that emphasized for purposes of the allocation of goods and services, free bargained-for exchange, rather than tradition or command. The heart of business contracts, as of the free market, is the bargain. Bargain, from the base-line of a status-organized society,

> is a tool of change and of growing individual self-determination, as is also any property regime which by increasing individual control increases the scope of experiment, the differentiation of holdings, and the factual effectiveness of the bargains of the wealthy. . .[These] bargain-effects. . .wax as bargains come to cover the future, and as they become in that aspect enforceable even though the other party breaks faith. [Moreover,] the self-determination aspect varies not only with the *number* of bargains which are in fact available to a particular bargainer, but also and most vigorously with the degree to which he has the wherewithal to individualize the phases of the bargain to his desires—, or, as the case may be, to sub-divide a single situation into a variety of specialized bargains to meet his needs. The power to shift one's status-in-block a single time (becoming a priest) or even often (marriage and divorce . . .) gets one a vast first step along the road "from status to contract."[9]

With increasing division of labor and with the free market as the mechanism for allocating resources, entrepreneurs have needed to bargain for the future delivery of resources and to rely on promises of future delivery in order to assure a continued flow of goods and services.[10] Contract law in the business sphere functioned in large part to ensure that bargains would be honored and legitimate expectations embodied in contracts respected. Contract liability, then, essentially is promissory liability. And in an industrializing, commercializing society, where wealth increasingly is comprised of promises, contracts—like other institutions of credit—played a vital role. In the spectrum between complete reliance for the satisfaction of one's business demands upon property and units owned and vertically integrated by oneself, on the one hand, and complete reliance on the "general spot market" (for example, flagging a cab), on the other, contracts involving future obligations lie somewhere in the middle. As such, contracts have contributed to the security and flexibility of relations between increasingly interdependent, specialized units under different ownership. In this context, protection of the interest of the individual promisee came to be seen as in the interest of the entire society.

In a rapidly changing economy, in which one wishes to maximize the effectiveness of bargaining as an efficient device, contract must be as pliable and responsive to the needs of the entrepreneur as possible. Freedom of contract was extended as to form by reducing ceremony to the irreducible minimum. And freedom of contract was extended as to substance by allowing the parties to tailor their private regimes to fit their special needs, by allowing maximum freedom as to the content of their contract. Thereby, "free" contracts enhance "the mobility of factors of production in the interest of the enterpriser who wishes to secure them in the most efficient way, so as to be able rationally to experiment with new methods of satisfying wants."[11]

These considerations of expediency were buttressed by a well-developed liberal ideology which enshrined freedom of contract in the temples of nineteenth and early twentieth-century capitalist society.[12] Although freedom of contract was adopted as a means to abolish feudal institutions, it came to be seen as an end in itself. At the core of its developed ideology was the notion made famous by Henry Sumner Maine that "the movement of the progressive societies [in world history] has hitherto been a movement *from Status to Contract.*"[13] Even more grandly, the free market and freedom of contract were associated with progress, envisaged in terms of social Darwinism as the survival of the fittest through the process of free competition. The related philosophy of government was, that government is best which governs least. Within this *laissez-faire* philosophy one of the state's few proper functions was to enforce by law contractual obligations. Viewed as incorporating individual volition and self-determination, contract came to be regarded as a central institution of liberal democracy, both in the economic and the political spheres (the latter in terms of social contract theory and constitution making). Dignified by individualistic utilitarianism which postulates that the individual is best able to maximize his own utility, by classical economic theory which holds that the good of society will be maximized by the interaction in the free market of individuals maximizing their own utilities, and by subsequent sociological theory which increasingly identified the expanded role of contracts with the growth of rationality, contracts came to symbolize liberty, democracy, and reason. Contracts, in the various nineteenth-century theories of the unilinear development of history, seemed likely to be one of the final victors of history.[14]

But this dominant vision of contracts always had been, and increasingly was, opposed by counter concerns. There is in the first place a basic contradiction or paradox in the dominant liberal conception of freedom of contract, which implies, even at the theoretical level where nineteenth-century Anglo-American jurists reveled, the need for limits on freedom in the very name of freedom. A contract to become a slave, it generally was agreed, is invalid: "The principle of freedom cannot require that [one] should be free not to be free. *It is not freedom to be allowed to alienate [one's] freedom.*"[15] Still, it is one thing to recognize theoretically the exception of the extreme case, and quite another to apply systematically the more subversive comprehensive notion, which gradually undermined abstract liberty of contract, that "necessitous men are not, truly speaking, free men, but, to answer a present exigency, will submit to any terms that the crafty may impose upon them."[16]

Limits on "Freedom"

Nevertheless, partly in recognition of this need to limit freedom in order to preserve it, and well before the twentieth-century shift in America to greater concern with freedom in fact as distinguished from abstract freedom at law, contract law even in its classic phase was kept from becoming purely abstract. Some of the limits imposed on abstraction can be explained within the framework of freedom of contract itself. For example, it is only common sense that the courts be sufficiently "particularistic" to preclude infants and the insane from contracting because of their "natural incapacity." Similarly, since the near exclusive concern of the institution of contracts was with marketplace-type business bargains, the courts' tendency to refuse the enforcement of gratuitous or one-sided promises frequently did not conflict with the spirit of freedom of contract.

Other derogations of abstraction, though, were more obviously based on public-policy determinations, which, however rationalized by the courts, flew in the teeth of freedom of contract. Notwithstanding, for example, the aforementioned classical statement of the doctrine of consideration, which enjoined the judiciary from evaluating the adequacy of a bargain, the courts have looked to adequacy as evidence of misrepresentation and fraud. Courts also have refused to enforce penalties. They have granted debtors rights of redemption not provided for in the contract, and even in the face of express contractual provisions to the contrary. Moreover, common law courts for reasons of public policy generally have refused to enforce contracts to commit crimes, torts, or immoral acts. Usury laws have limited enforceable interest rates.

But if abstract freedom of contract never was realized in full at common law—contrary to the picture the famous ideological, American freedom of contract cases of the late nineteenth and early twentieth centuries presented as they rewrote history[17]—the late nineteenth and twentieth centuries saw new and growing restrictions on abstraction in the form of legislation. This development was part of a broader socio-economic trend "to protect the weak against the strong, the farmer against the railroad, the laborer against the employer, the borrower against the lender, and so on.... [N]ew rules... [were] born, and to that extent ['pure'] contract...[was] displaced."[18] In the twentieth century the legislatures in the name of the public interest increasingly came to assert control over contracts.

Dual Nature of Contracts

Contracts in the West can be thought of in terms of their dual nature, involving legal freedom for the individual parties and the exercise of state authority or control. These two factors, freedom and authority, are necessary and characteristic elements of the institution of contracts as the West understands it. These factors coexist in varying relationships in all phases of contracts. Their interaction may be thought of in terms of a spectrum with the poles of pure freedom and pure authority at either end. Given this perspective, agreements of purely polar character between the parties—those lacking one of these factors—cannot be considered contracts.

The polarization of freedom and authority is clearest in the realm wherein contracts are held to be absolutely invalid by the state. Here the individual is not free to bind himself. Here the interest of the society is considered so vital by the society that it asserts its authority in a categorical manner to preclude a binding effect. The concern for abstraction is sacrificed, and in a sense, the absolutely invalid "contract" is not a contract at all.

At any point along the continuum from freedom to authority, however, this polarity is less clear. The element of authority must be present at least to a minimal degree as a means of facilitation—a means of setting the framework within which private lawmaking can function—and as the structure for enforceability, the structure that encourages voluntary performance of certain promises and provides redress for breach thereof. On the other hand, the element of minimal freedom also must be present to distinguish contractual obligations from those obligations the society imposes by fiat directly on the individual or juristic entity. Where the element of legal freedom is lacking, the basic nature of the institution of contracts as a mode of private lawmaking between the parties is denied. In such circumstances the institution becomes another device in the galaxy of devices for direct societal intervention and ordering in a sphere that the Western classical liberal tradition reserved to private initiative.

Nevertheless, although contract has a dual nature and although freedom of contract in the West increasingly has been limited at law, the "residuary" view of contract highlights the significance of freedom and abstractness for contracts and explains much about the rigidification of Anglo-American contract theory since the nineteenth century. The extent to which, during the past century of revolutionary change in the economy and far-reaching change in the content and orientation of the Anglo-American legal system, the rules of contract law—that body of law most intimately connected with the developing capitalist economy—have remained static might at first seem surprising. Seen as a residuary, covering the "residue of economic behavior left unregulated (the free market)," however, the static quality of contract rules is natural, in fact definitional. Changes in the *scope* and *impact* of pure contract law, *rather than its internal rules,* reflect the fantastic changes in the economy since 1850.

> By definition, no revolution could take place [in contract rules] because contract law acted as a residual category, its content determined mainly by what law did in other respects affecting economic behavior. Instead, types of transactions marched in and out of the area of contract. In the early part of the Nineteenth Century, the law of contract grew fat with the spoils of other fields. Land. . .,leases, claims, rights, and intangible interests became subject to bargain and sale; . . . free labor replaced slave and bound labor . . . Contract law expanded and narrowed its applicability to human affairs primarily through a process of inclusion and exclusion. . . Major exceptions to. . .[t]he basic rule of contract law. . .that contracts are presumptively valid and enforceable according to their terms. . .[were] changes in the effective content and reach of the law of contract, but they do not look in form like changes in the law of contract; they look like labor law, insurance law, social security or public utility regulation. . . [Thus,] the most dramatic changes touching the significance of contract law in modern life also came not through internal developments in contract law, but through developments in public policy which systematically robbed contract of its subject matter.[19]

Therefore, the archaic court-made rules of contract law, systematized in treatises, continue to stress abstraction in its pristine purity. For legal scholars trained in American law schools through the case study method—essentially the study of appellate court decisions—this pristine approach, which dominates the segregated law course on contracts, is likely to dominate their comparative analysis. Despite the usual perfunctory qualifications regarding the operational reality of contractual freedom in the West, comparative analysis of contracts frequently begins, and too often ends, bound by the biases of court-developed contract theory. At its worst such analysis can be "compared to a zoology course which confined its study to dodos and unicorns, to beasts rare or long dead and beasts that never lived."[20] It is difficult to break these confines, as the next three chapters, turning the analysis to China, in part attest.

Chapter II

CONTRACTS AND PLANNING IN CHINA, 1949–1963
AN OVERVIEW

The next three chapters are concerned with the nature, functions, and scope of the institution in the People's Republic of China that is commonly referred to as contracts. These chapters as a whole ultimately focus on three problems: (1) the nature of the institution; (2) whether the institution, if it is of a different nature from the Western contract, performs tasks in the Chinese context not unlike those of the institution of contracts in the West; and (3) how it performs those tasks.

Chapter II deals with the general ideological, historical, and economic context in which the institution has operated in China. It also describes its chronological development. And, because of the crucial impact of the planning structure in Communist China on contracts, a lengthy description of the planning structure is included. The strictly legal aspects of the institution of contracts are not apparent at the level of discussion of this chapter. Chapter III, therefore, discusses the legal rules governing the operation of this institution and attempts to draw implications regarding its nature, functions, and scope from those rules by means of comparative and functional analysis. Chapter IV is a case study of contracts in action.

The institution of contracts that developed in Communist China was shaped by the ideological, historical, and economic context in which it grew. On the ideological plane, the theories of Marx, Lenin, and Stalin molded the Chinese Communists' general understanding of law. On the historical plane, the Russian model of a planned economy was accepted initially as the model for the development of a socialist society. And on the economic plane, that model, functioning under Chinese conditions, molded the course of development of the institution of contracts in China. It is within this framework of economic planning that the future of contracts in China lies.

The Ideological Context

The ideological lineage of Chinese communism with regard to law can be traced from classical Marxism through the Russian experience. According to classical Marxism, law is an historical phenomenon that reflects the economic relations between classes at a point in history. Law changes as the economic relations change from stage to stage throughout history. The state and law are parts of the "super-structures" built on the economic base. Law, as an adjunct of the state, develops with the appearance of classes and the assertion of dominance by one class over another. Thus, law is a tool for class oppression, a tool of the ruling class in the class struggle. As such, it will "wither away" and disappear in the classless, stateless Communist utopia of the future.[1]

Classical Marxism offers a "critique of law."[2] It predicts that law must ultimately disappear, but it does not explain how the transition from revolution to utopia is to be made without law.[3] The Soviet experience in the pre-1920 period and in the early 1930s illustrates this failing. In each instance an attempt to denigrate and eliminate law in compliance with the tenets of classical Marxism resulted in a restoration of law.[4]

In the high tide of revolutionary zeal following 1917 all legal codes in Russia were abrogated. This approach, however, proved disastrous, and with the institution of Lenin's New Economic Policy, 1921–1928, law was partially restored. Codes were promulgated to facilitate the workings of the mixed capitalist-socialist economy that was set up. While this return to bourgeois law was said to be, and in fact was, only temporary, it nevertheless represented a recognition of the need for a system of law. It also reflected the inadequacy of classical Marxism in this regard, an inadequacy that was noted by a Russian jurist who said: "Law is that sphere of social relationships to which Marxist thought has turned its attention less than to all other aspects of social relationships."[5]

Notwithstanding its partial restoration in practice at this time in Soviet Russia, law did not attain acceptance at the ideological level. The views advanced by Pashukanis—that law achieved its full flower under capitalism, that all law is inherently bourgeois, and that law even in a socialist society can be no different—prevailed until the mid-1930s.[6] Law was derided as a "'bourgeois fetish' doomed to disappear with the introduction of socialism."[7]

When the First Soviet Five-Year Plan was put into effect in 1928, the continuing view that law had no place in a socialist society was manifested in the notion that the conservatism of law would be replaced by the dynamism of plan.[8] Laws that temporarily remained in existence were subordinated to Party and Plan.[9] But this contempt for law, which had shown itself in the period immediately after 1917 only to be rejected in practice by the New Economic Policy, was again to be rejected, this time in effect at the ideological level as well as in practice.

By the mid-1930s the general deterioration of the legal system, resulting in part from this contemptuous view of law, had adversely affected production as well as the society at large.[10] As a consequence, Stalin in 1936 moved to restore legality,[11] at least in the nonpolitical realm.[12] Recognizing the impracticality of the earlier approach, Stalin declared, "We need stability of laws now more than ever."[13] Through Vyshinsky, Stalin indicated that the ultimate "withering away" of state and law was not imminent,[14] and that in fact state and law had been reconstructed on a new socialist basis.[15] "Socialist legality" became a positive good rather than a mere expediency.[16]

This renewed emphasis on legality was motivated by the desire to stabilize the regime and to facilitate planning.[17] The movement for "socialist legality" was restricted to nonpolitical areas in which the regime felt secure.[18] Still, it is one of the most significant internal developments in Russia since 1917. Moreover, this movement has continued to the present.[19]

Nonetheless, acceptance of the role of law is not equivalent to acceptance of the rule of law.[20] Law for the Soviet regime is one means of social control, not the sole or primary means of control.[21] Law ultimately is at the service of the Party and, whenever the Party feels the social order it has created is threatened, it may reject law and rule by fiat. However, although law on the one hand is ultimately subordinated to nonlegal methods, it is, on the other, relatively supreme where the regime does not feel threatened.[22]

This dualism is inherent in the notion of law held by the Chinese Communists. While it is beyond the scope of this book to discuss the current relevance and importance of the traditional Chinese concept of law, suffice it to repeat the truism that law was recognized by the Confucian bureaucracy as a means of social ordering but not as the primary or the best means.[23] In this respect the inherited traditional concept of the function of law prefigures the Soviet view adopted by the Chinese Communists. Though the principle of the rule of law has not been accepted by the Chinese Communists, neither has law, as a genre, been considered the ideological anathema it was in the Soviet Union prior to the mid-1930s.[24] The Chinese revered the category of socialist law at the founding of the regime in 1949. Although the Chinese recognize the class nature of law, they accept law as an instrument of the ruling worker-peasant classes, to be employed in the building of socialism and communism. Law, as a method of ordering and facilitating, appears to have a future in Communist China.

It is a basic postulate of the Marxist and Chinese Communist concept of law, and a fact of societal life, that the role of law in any period is determined by the particular economic and political conditions then existing.[25] The role of "contracts" in China has evolved and changed during three economic periods since 1949.

The Broad Economic Context and the Russian Model

At the founding of the People's Republic in October 1949, the Chinese Communists inherited the economic problems that had plagued China for centuries as well as those arising from the chaos in China during the first half of the twentieth century. The traditional problems such as population pressure on the land, fragmentation of farms, inadequate transportation, scarcity of capital, a subsistence standard of living, and poor flood control had not been met, and were hardly faced, during the first fifty years of the century. In addition, problems arising from the almost continuous state of war in China during the century—inflation, the general disruption of all economic systems, and the isolation of rural from urban areas—faced the Communists.[26]

At the outset the Chinese Communists decided to follow the broad outlines of the Russian model. From Lenin they took the idea of the New Economic Policy, a mixed economy consisting of state, state-private, cooperative, and private sectors. This policy, embraced in the program for the period of the New Democracy, was applicable to the early stage of development during which the private sector was needed.[27] From Stalin they took the concepts of overall economic planning, socialism in one country, and agricultural collectivism.[28] These were goals to be achieved during the second stage of the revolution after the period of the New Democracy.[29] The first stage essentially was to prepare for the second.

The First Period, 1949–1952

The Economic Context

The Common Program of the People's Republic,[30] adopted on September 29, 1949, by the Chinese People's Political Consultative Conference,[31] stated these goals and the timetable for the prospective regime. Articles 26–40 deal with economic policy.[32] They set forth the policy of the regime for the first stage and broadly define its ultimate goals. The state-owned sector is "the leading force of the entire . . . economy."[33] The cooperative sector, which is second in regard only to the state-owned sector, is to be accorded "preferential treatment."[34] Private enterprises "beneficial to the national welfare and the people's livelihood," however, are also to be encouraged and fostered by the government.[35] The central government under Article 33 is charged with the task of drawing up "as early as possible" a "general plan for rehabilitating and developing the main branches of public and private economy" and with the job of coordinating the entire economy.[36]

The design for the future emerges from the Common Program as a whole. The state sector is to be dominant. The private sector, which will be socialized gradually, is to be subject to the needs of the former. The development of heavy industry is to be given priority while agriculture will be cooperativized step-by-step. All will be under an economic plan drawn up by the central government.[37]

The Common Program is specifically applicable to the first period, the period of economic rehabilitation. During the first part of that period, from 1949 to mid-1950, the government focused on stabilizing conditions. Through such devices as state control of the wholesale trade, the wage-point system, the

public supply system, and currency control, the government stemmed inflation. It used the People's Army to rebuild devastated transportation systems,
thereby improving communication between rural and urban areas.[38]

During the later part of this period of rehabilitation the government
concentrated on extending the area of state control over the private sector of
the economy and on preparing for the First Five-Year Plan.[39] Through such
campaigns as the "Five-Anti movement," which was aimed at intransigence
and economic crimes against the state, the regime extended its control over
the private sector.[40] Concurrently it moved to improve efficiency and eliminate inadequacies in government.[41]

Planning, 1949–1952

At this stage planning was in its embryonic form. Due to the lack of statistical information, due to the importance of the private sector in industry
and agriculture—a sector that is difficult to integrate into an economic plan—
and due to the lack of skilled personnel and of experience in planning, the
central government contented itself with a number of partial plans primarily
regulating the state sector of the economy.[42] It also set up a crude planning
structure, held planning conferences, and set production targets for critical
goods.[43] There was, however, no overall plan.[44]

Prior to November 1952, formal responsibility for planning lay in the
Planning Board of the Committee for Financial and Economic Affairs
(CFEA).[45] In the governmental hierarchy, the CFEA was directly under the
highest executive organ, the Government Affairs Council.[46] Below the CFEA
of the central government were similar committees at the lower levels of government[47] and planning sections and bureaus in various agencies at all levels
of government.[48] While the CFEA was charged with the duty of formulating
an overall plan, it in fact merely integrated estimates made at the ministerial
level.[49]

In November 1952, on the eve of the announcement of the First Five-
Year Plan, and in recognition of the fact that a more efficient and comprehensive planning structure was needed for the next stage, the State Planning Committee[50] supplanted the Planning Board of the CFEA.[51]

By this time the basic method for formulating plans had been developed.
Under this method each planning organ in the hierarchy, based on information culled from its immediately subordinate organs, sends to the latter provisional targets known as control figures. On the basis of these control figures, which may relate to such different aspects of production as volume,
cost, and supply, the subordinate organs draw up their plans relating to each
of these aspects and send them up to their superior. Integrating these plans,
the highest planning organ makes a final "balance" based on such factors as

resources available and the priority schedule of goals for the current period. These determinations, which ultimately must be approved by the National People's Congress, are then promulgated as the Plan.[52]

As part of the preparation for overall economic planning the government in 1951 attempted to introduce the business-accountability system in enterprises under government control.[53] In these early years, though, the system was not effectively enforced by the regime.[54] The business-accountability system, like economic planning in general of which it is a component part, was transplanted from Russia. Russia in the initial years of its First Five-Year Plan had attempted to implement the plan exclusively by administrative fiat without recourse to civil law institutions. But in the early 1930s to improve efficiency Stalin began to emphasize the importance of the business-accountability system and its correlative, the civil law contract.[55] China, profiting from this example, appears to have been aware of the need for the business-accountability system in the period before overall economic planning.

The system of business accountability is said to be the application of normal principles of good business to state enterprises.[56] It is predicated upon the notion that these enterprises are to be treated in certain respects as if they were independent businesses operating for a profit. Under the system, an enterprise's rate of return is to be determined on the basis of prices often arbitrarily set by the state.[57] The success of an enterprise is measured in part by its rate of return. Each enterprise is allocated fixed assets and liquid funds to be managed by the enterprise in accordance with the plans. Each is to treat itself as an accounting unit for purposes of comprehensive cost accounting and financial reporting. Moreover, every enterprise is to draw up its own more detailed plans on the basis of state plans. Each is to be managed efficiently by a central plant manager who is a state official. Each is to have its own bank account in the state bank.[58]

Under the system incentives are given to management and workers to fulfill and overfulfill state plans. These incentives are crucial to the system itself as it is only through the pecuniary interest of the workers involved that the system can be properly applied.[59]

In an economy in which capital is scarce, the primary purpose of the system is to increase efficiency, to maximize the use of resources, and to encourage maximum capital accumulation by each enterprise while fulfilling its plan. It has been said in *Hsueh-hsi* (Study) that:

> The central theme of the business accounting system is reduction of production cost under the standing requirement of the quality of products being maintained. The operation of the system demands that the enterprises concerned should consume the minimum amount of materials and labor in the fulfillment of their designated tasks. It demands

that the enterprises themselves depend on proceeds from the sale of their products to defray all production and sales expenses, leaving at the same time a margin of profit. This is the essence of the business account-ability system.[60]

By means of requirements for record keeping and financial reporting, by means of the allocation of responsibility to particular units, and by means of material rewards incorporated in the system, the system of business account-ability functions both as a measure of performance and as an incentive to performance.

Contracts, 1949–1952

It is first in connection with the allocation of responsibility within an enterprise and between enterprises that contracts in this period and increas-ingly since then have functioned. Within an enterprise contracts were con-cluded to divide responsibility between various production units and to co-ordinate more carefully the work of these units. Contracts also were used to make targets in the plans of the enterprises more concrete to the individual worker. In this way the system of business accountability and the crude plans were to be implemented.[61]

Use of contracts between enterprises was advocated for similar reasons. In order to keep accurate accounts it is necessary to determine where, how, and why disruptions and mistakes occur. Contracts between enterprises, spec-ifying the details for performance under plans, can serve as a means to obtain such information. By litigation of specific disputes between enterprises, re-sponsibility for failures can be placed.

At this early period it was in part for these reasons that such contracts were proposed.[62] While it is not clear to what extent these proposals were carried out in light of the lack of skilled personnel and the unrefined, pre-natal development both of planning and of the system of business accountability, it appears unlikely they were implemented to a significant degree at this time.[63]

During this period, and thereafter until 1956 by which time the private sector of industry had been substantially incorporated into the state sector, however, the institution of contracts played another more significant role. In line with the dual desire of the state to bring the private sector gradually un-der control of the central government and at the same time to utilize that sec-tor in the production process, government agencies began awarding purchasing and processing contracts to private enterprises. The government offers to con-tract, which could not be refused by private enterpreneurs without offending the regime and "the people," once accepted, served as a wedge by which state control gradually reached out into the private sector. Although at the outset in early 1950 it appeared that this contract letting simply was aimed at assist-ing the rehabilitation of private production, it became clear by late 1950 that it had another purpose. Through contracts the government began to direct the development of private industry.[64] Contracts of this sort, in conjunction with government control of the distribution of raw materials and of credit and marketing facilities, provided an additional means to stimulate the transformation of private industry into socialist industry.[65]

Once in a position of direct influence, the government encouraged concentration within the private sector. It pressured private enterprises to transform themselves into joint state-private enterprises. Step-by-step it extended to the semi-private sector regulations that previously had applied only to the state sector. The stage was set for the transition to the next period, the period of overall economic planning.

The Second Period, 1953–1957

The Economic Context

The second period began under conditions not generically different from those existing in 1949. While a large measure of stabilization and rehabilitation had been effected during the first period, problems such as inflation and lack of communications between urban and rural areas had only been mitigated, not substantially solved. More importantly, many of the deeper problems that had existed in China for centuries[66] still remained to be faced. During this period the regime undertook to face and to begin to solve these problems.

The regime's methods were essentially those announced previously in the Common Program[67] and affirmed subsequently in the Constitution of the People's Republic[68]: overall planning, emphasis on rapid industrialization, and socialization of the private sector of the economy.[69]

The application of these methods in 1953, however, presented two basic problems. First, unlike Russia at the time it introduced its First Five-Year Plan, a major part of China's industrial economy was still within the private sector when the Chinese government announced the beginning of its First Five-Year Plan in late 1952.[70] This meant that planning would be more difficult, as control over the private sector was still limited. In light of this difficulty, China's decision to deviate from the Russian experience and to initiate the First Five-Year Plan before it had secured effective control over the entire economy may reflect the government's eagerness to embark on planning.[71] This deviation may also reflect a realistic recognition of the second problem the government faced.

At the time of the announcement of the First Five-Year Plan, China lacked the personnel, planning structure, and technical experience necessary to carry out overall economic planning. It was only through trial and error in actual planning that the regime developed the capacity to plan. Not until mid-1955, two and one-half years after the official beginning of the plan, was the regime prepared to publish the details of the plan. Thus, a major portion of the planning effort during the first half of the plan period was in fact consciously devoted to gaining experience in planning and to setting up the basic structure for formulating and enforcing the plan.[72]

Planning, 1953–1957

During the First Five-Year Plan the basic planning structure was in flux.[73] The trend toward centralization, noticeable in the first period, continued until 1957.[74] The governmental structure that developed under the State Council was complex. It consisted of a multitude of interacting and interlocking hierarchies all ultimately responsible to the State Council.[75]

Directly below the State Council at the central government level are the State Planning Commission, the National Economic Commission, the State Statistical Bureau, the National Technological Commission, the General Supply Bureau, the various ministries, and the People's Bank system. In addition, a National Construction Commission existed under the State Council from 1954 to 1958.

For purposes of this discussion, the basic tiers of government below the central government level can be taken to be the provincial level, the county level, and the village level.[76] At each of these lower levels a system corresponding to the executive system of the central level exists. Thus, at each level there is a people's council corresponding to the State Council and departments corresponding to the ministries. Each level is subordinated to the authority of the level above it.[77]

At the center of the governmental planning structure is the State Planning Commission, created in 1952 and brought under the State Council in the reorganization of 1954.[78] This body, until mid-1956 when the National Economic Commission was set up to formulate and direct annual planning, had primary responsibility for the formulation and implementation of long-term and annual plans. Subsequent to that date the State Planning Commission has concentrated on long-term planning.[79] Directly subordinate to the commission are the planning departments in the ministries of the central government. Below the latter are the planning bureaus at the provincial level, and below them the planning units in the local departments corresponding to the ministries.[80] The lowest level in the chain of command is the planning unit in the individual business enterprise.[81]

The National Economic Commission fits into this same hierarchy. It works closely with the State Planning Commission in drawing up annual plans.[82] The State Statistical Bureau, set up in August 1952 directly under the State Council,[83] is charged with the task of collecting, coordinating, and providing statistics. These statistics are vital to the performance of planning functions by the State Planning Commission and the National Economic Commission.[84] Below the Statistical Bureau are statistical sections that are part of the planning units at each level.[85]

As part of the trend toward specialization of functions manifested in the proliferation of ministries and in the subsequent creation of the National Economic Commission, the National Construction Commission was created in November 1954. Its purpose generally was to control capital construction under the First Five-Year Plan and specifically to coordinate key projects built with Russian aid. This commission apparently also had its counterparts at lower levels.[86] In early 1958 the commission was abolished, and its functions apparently taken over by the State Planning Commission, the National Economic Commission, and the ministries concerned with capital construction.[87]

Little is known of the General Supply Bureau and the National Technological Commission. The latter was established in May 1956 to facilitate and coordinate the implementation of new technology.[88] The General Supply Bureau, created at the same time, has the job of insuring the smooth flow of materials, a flow that is crucial to the realization of planned targets.[89]

As of mid-1955, at least twenty ministries appear to have been directly concerned with industrial administration and planning.[90] Policy determinations of the ministries are carried out by the aforementioned subordinate counterparts at the various tiers. In addition, specialized industrial companies have been set up under the ministries[91] to give the latter greater central control over enterprises in the particular industry regulated by the companies. These specialized industrial companies served to stimulate the process of socialization within a particular industry,[92] and, most significantly for present purposes, they may also have served to resolve contract disputes between enterprises under their jurisdiction.[93]

Of the ministries, the Ministry of Supervision has played an especially important role since 1953 in enforcing the plan at all levels. By means of on-site inspections and the elaborate network of supervisory sections at ministry, bureau, and enterprise levels, it works to maximize efficiency in implementing the plan.[94]

The centralized system of People's Banks with local branches also fulfills an important supervisory function. Since state-owned enterprises keep their accounts with the state banks and since transfer of funds on the accounts of the banks is the primary method for an enterprise to satisfy its debts, banks can and do supervise the contracts from which these debts arise. Moreover, the banks demand periodic financial reports from the financial control offices in the enterprises.[95]

This is a simplified view of the formal governmental planning structure. This structure is supplemented by and ultimately subordinate to a parallel and overlapping Party hierarchy. The focus of ultimate power in China has been in the hands of the Politburo of the Party and its executive, the Secretariat. The broad outlines of planning policy are drawn by the Politburo. The State Planning Commission and other governmental organs then act within these outlines.[96] Similarly, final authority to direct the execution of the plan rests with the Secretariat of the Party which exercises that authority through the Party organization that reaches to all levels and units of government and industry.[97] It is an understatement to say that Party control is considered crucial to the success of economic planning.

It is not clear how this exceedingly complex and bureaucratic system functions in detail. The manner in which it operates has been suggested above in general terms. For purposes of this chapter it is important to note four further points with respect to planning: (1) planning does not and cannot cover all products; (2) even as to the products it does cover, the degree of specificity regarding essentials and details of performance varies from product to product; (3) plans issued by the central government are not broken down into time periods of less than one year; and (4) plans subsequently broken down by lower echelons often are not broken down to the smallest units of production. Consequently, there is an area not explicitly dealt with by the plan proper in which individual initiative is allowed and necessary. Moreover, plans themselves are hardly self-executing. It is in these spheres that contracts have played an increasingly vital role. Whether the "freedom" allowed the parties in these spheres is sufficient to satisfy the minimal level required by the given definition of contracts (in the Western sense) remains to be seen.

Contracts, 1953–1957

Contracts during this period continued to be used both to extend governmental control over private enterprises and to coordinate activities within the state sector and between the state and other sectors. Their role during this the second period seems to have been increased.

Since the socialization of the private sector of industry was not achieved until 1956, contracts prior to that time continued to be used to exploit private industry and to speed its transformation.[98] After that time the surviving private enterprises were transformed into state-private joint enterprises. Then the private entrepreneurs were no longer entitled to a share of the profits arising from state contracts but were limited to a fixed rate of interest on their original investments, a rate based directly upon government regulations rather than contract.[99]

Contracts probably continued to be employed in transactions between state enterprises and internally between various units within an enterprise.[100] While no mention has been found of the use of general contracts as a framework for subcontracts, it is likely that general contracts between ministries did form the framework for subcontracts between enterprises.[101]

Contracts, in addition to assigning responsibility between parties, probably were utilized to define the plan further so that it could be implemented smoothly. As planning was expanded to cover more products and more units, this coordinating function of contracts became more important. The more tightly knit the plan became, the more disastrous was a failure to realize it on any level of the system. Coordination was essential.

During this period contracts also began to play a more significant role in the relations between state commercial organs, on the one hand, and individual peasants and incipient peasant collectives, on the other. In order to achieve the goals in planned purchase and planned supply the state commercial organs executed sales contracts with individuals and groups. On the basis of these contracts for purchase of goods by the state agencies, advances were made to the peasants that provided them with a means of subsistence and the means to purchase seeds and such needed for production. The state, correspondingly, was assured by these contracts of its supply of agricultural products.[102]

Finally, it appears that contracts were concluded between various kinds of cooperatives. Support for this practice is reflected in Mao Tse-tung's words in 1956: "The conclusion of . . . contracts between supply and marketing cooperatives and agricultural producer cooperatives should be extended universally."[103]

The Third Period, 1958–1963

The Economic Context

As the First Five-Year Plan drew to a close in 1957, it became apparent that, while it would be overfulfilled in many respects, changes were in order. The extreme degree of centralization, the emphasis on large-scale modern industry, and the relative inattention to agriculture had produced grave problems. Centralization had resulted in inflexible planning that failed to take account of the great differences between localities. Moreover, it had hampered local initiative and correspondingly dampened local enthusiasm. The emphasis on large-scale modern industry had discouraged the development of small-scale, labor-intensive, local industry. As a consequence, the high demand for capital necessary to develop modern plans added to the already severe inflationary pressures within the economy—pressures that had been produced initially in large part by the inattention to agricultural and consumer goods.[104]

Recognition of these problems was apparent in the government's easing of the pace and broadening of the focus of development immediately following the severe inflation in 1956. [105] In 1957 the government conceded that agricultural and consumer goods would have to be given greater attention in the Second Five-Year Plan. During the year the government stressed the need to exploit local initiative.[106] Then, in late 1957 the decision was made to decentralize planning and control of numerous enterprises. Local governments were to play a more important role in both.[107] This crucial decision to decentralize prepared the way for the Great Leap Forward.

In the beginning of 1958, the first year of the third period, the regime called for total mobilization of the economy for the Great Leap Forward. The caution of 1957 was rejected. Mass movements were organized to stimulate production. Targets were raised and raised again.[108] The country was to "walk on two legs": not only heavy, centralized, modern industry but also light, localized, and native industry was to be encouraged; not only industry but also agriculture would be developed; and central leadership would be supplemented by increased local management.[109]

In mid-1958 the commune movement in agriculture was announced. The peasants, many of whom had only recently obtained their land in the land reforms during the 1949–1952 period, had been increasingly cooperativized during 1955–1956. This drive toward agricultural concentration and co-operativization culminated in the commune movement whereby large numbers of peasants were placed under a single local administration that managed all facets of life and production within its confines.[110] Communization, it was said, would facilitate the Great Leap by releasing and more efficiently coordinating untapped labor resources and by realizing economies of scale.

The year 1958, the first of the Second Five-Year Plan, produced increases in industrial production. But in the frenzy of the campaigns it also produced exaggerated claims, poor planning, and low-quality products. Furthermore, communization decreased material incentives and caused resentment. After the high point of the Great Leap in late 1958 the regime beat a strategic retreat. The communes, in effect, were stripped of much of their concentrated power, and individual exertion was encouraged through greater material incentives.[111] The regime since 1958 has acknowledged the need for better statistical information and planning and criticized the exaggerated and shoddy reporting during 1958 and 1959.[112] In general it has recognized that one must look before one leaps.

Planning, 1958–1963

Decentralization and the Great Leap had a strong impact on economic planning. Under decentralization of planning and management the number of "absolute" control figures handed down by the central government was reduced from twelve to four. Control figures for the other eight categories rather than being "absolute" were made "relative" or merely "consultative."[113] Thus, the lower tier planning units were given greater latitude in setting their own targets for categories not covered by "absolute" norms. The freedom granted, however, appears to have been generally limited to the freedom to choose methods of fulfilling planned tasks rather than to determine the tasks themselves. Moreover, this decentralization of planning may not significantly have affected contractual freedom, since decentralization largely meant handing planning responsibility from higher to lower level government organs.

After decentralization, plans drawn up by local planning units covered enterprises formerly under the direct administration of the central government as well as enterprises previously managed by local authorities. Decentralization was a recognition of the "unbalanced economic development in different regions."[114] Local units were empowered to draw up "balances" for all enterprises in their area in order better to coordinate the supply and production of the area as a whole.

The policy slogan for the period was that the national economy should be taken as a "coordinated chess game."[115] This meant that local governments in formulating their plans should respect the interests of the nation and that "as a rule, larger interests should determine smaller interests, smaller interests should yield to larger interests."[116] Hence, the delegation of added planning powers to local governments was predicated on the requirement that local governments subordinate narrow interests to national interests. The delegation reflected a desire to stimulate local initiative and to take local conditions into account but did not represent any affirmative willingness on the part of the government to decrease its control. In fact, although the movement was characterized by decentralization at the national level, it was also characterized by a concentration of power at the local level. Thus, far from decreasing government control, it gave government new means for control. Nevertheless, the above changes were sensible in terms of efficiency. The pattern established appears to have been perpetuated into the 1960s.

The excesses of 1958 subsequently were corrected. In 1958 the regime had instituted a new method of formulating plans under which the escalation of targets was built into the means of formulation and reporting. This method in large part was responsible for the exaggerated claims regarding production in 1958.[117] In addition, during that year the statistical system was allowed to deteriorate with the result that the restraining influence of that system upon falsification and exaggeration was removed.[118] While it is not clear whether the escalation method of planning was still in effect during 1959, it was not stressed after 1958.[119] During 1959 and 1960 the state statistical system was rehabilitated and, in fact, more firmly established than before.[120]

Contracts, 1958–1963

In the context of this period of decentralization, religiously frenzied movements, and increased local concentration in agriculture, the necessity for concluding contracts was repeatedly emphasized. The importance of concluding contracts to the realization of the second plan was particularly stressed in the two areas of capital construction and commune agricultural production.

A resolution of the second session of the Eighth Party Congress in 1958 stated that "the contract system should be extensively introduced between the state and communes and between different communes."[121] A lengthy article in *Ta kung pao* underlined this point.[122] Contracts, it said, were needed to: (1) "[draw] a number of major products [produced by the communes] into the orbit of the state plan";[123] (2) further develop specialization and diversification of production through direct exchanges between communes of products not within state plans; (3) "link up the state plan closely with plans of the rural people's communes";[124] (4) solve specific problems with respect to products covered by state plans; (5) provide a framework for advance payment to the peasants; (6) stimulate production of sideline commodities and consumer goods; (7) strengthen the worker-peasant alliance, that is, facilitate the movement of products between rural and urban areas; (8) mobilize the masses; (9) insure that targets set will be flexible and responsive to local conditions; and (10) educate the peasant to understand planned leadership.

In the light of the virtual elimination of private ownership and of free markets in 1958–1959, new means of insuring production and exchange of goods on a large scale were needed. Since state plans for the commune only prescribed in detail the targets of major products, the commune had substantial latitude in drawing up its plans even as to some products mentioned in the state plans. Moreover, since state plans of necessity could not deal with many minor products at all, the production of these goods in effect was left to the discretion of the commune. Products that were not included in state plans, therefore, might not be produced unless other means were available to encourage their production. Contracts between state commercial organs and communes provided one means of encouragement and also served to direct commune planning toward state needs.[125]

In the late fifties and early sixties an increase in the production of sideline agricultural and light industrial products was felt by the regime to be vital to satisfy the dangerously high demand for consumer goods. As before, contracts were useful in stimulating this increase by providing the producer in advance with an assured market for goods to be produced, by giving the producer an assured supply of raw materials needed for production, and often by making available to the producer advance payments that could be utilized for production.[126]

Establishing large communes not only created an increased need for methods of channeling goods on a large scale throughout the country, but it also made the task of concluding contracts easier as the number of units with which the state commercial organs had to deal directly was reduced. Concomitantly, as the unit dealt with became larger, there was a greater need for allocation of responsibility within the unit. Contracts served in this latter capacity as they had in earlier periods.

Tasks assigned to production brigades and production teams within the commune were incorporated in contracts between the commune and the brigades and between the brigade and the teams. These contracts were formulated at conferences led by party members at each level. The aim of the conferences was to set appropriate targets and to stir up mass enthusiasm. The discussions and allocation of responsibility down to small units were thought to give the individual a greater feeling of participation and to stimulate mass initiative in adopting methods to fulfill contracted tasks in the light of local conditions.[127]

The contracts were said to provide built-in incentives to overfulfillment, because contract targets were supposed to be set sufficiently low so that under normal conditions overfulfillment is probable. And "any yield over and above the contracted amount would belong almost entirely to the commune members of the production unit."[128] Moreover, the discussion of comparative performances by other production teams and brigades, which was a part of the procedure for drawing up contract targets, served as a basis for emulation drives. Material rewards for good performance in these drives supplemented incentives built into the contracts. Contracts thus became a means to make the overall task of the commune concrete for the individual and to stimulate him to achieve his particular task.[129]

Although the communes internally have been decentralized since 1959 and although relatively free markets have appeared again,[130] it seems that the contract still plays a significant role in this area. In fact, while it may be more cumbersome to conclude contracts directly with the myriad production teams and brigades that now have greater autonomy, it is probably more important to do so, because it would be nearly impossible to integrate units so small and numerous directly into state planning. Moreover, in the light of famines and other shortages in the early 1960s all methods to stimulate local initiative are likely to have been utilized.

During this period contracts were also used extensively in the crucial area of capital construction. A system for using contracts in capital construction, said to have sprung from the spontaneous enthusiasm of the masses and later called the "investment-by-contract system," was decreed by the State Council in 1958. The system was further refined in supplementary provisions of the State Council subsequently published in May 1959.[131]

In essence these provisions give greater latitude to construction units to use, in a flexible manner, funds appropriated for a task.[132] The 1958 provisions provide that:

> On condition that production capacity [of the enterprise under construction] was not reduced, that completion of projects was not delayed, and that the total amount of money available [to the unit] for investment was not exceeded, the funds earmarked for investment in capital construction in the state plans and the state budget were turned over to the construction units concerned for use at their own discretion under the investment-by-contract system.[133]

The purpose of this aspect of the system was to eliminate the woodenness of detailed high level planning and to encourage ingenuity and economy in the light of particular local conditions.[134] Again, this represents a granting of expanded freedom to select means and the continued withholding of the freedom to determine goals.

In addition to the limited discretionary power given to the construction units in order to stimulate adaptation to local conditions, the State Council provisions also authorized the construction units to retain any funds saved by the unit's economizing, the funds saved to be used by the concerned units, with the approval of higher echelons, to further enlarge the scope of the construction project. The 1959 provisions emphasized the requirement that the discretion given the construction units was on condition that fulfillment of the plans be guaranteed.[135] These provisions abolish the previous rule that appropriations must be allocated for specific uses in each project and that "the funds for oil may not be used for vinegar."[136] They also abolish the rule that any surplus has to be returned directly to the state.[137]

Under this system contracts are to be concluded at all levels.[138] There are contracts horizontally between the construction unit, the enterprise under construction, and the designing unit. And within each unit concerned there are vertical contracts reaching down to the smallest component unit. Powers and responsibilities are delegated and plans are formulated by the component units on the basis of the targets those units contracted for.

The system of "investment-by-contract" is said by *Ts'ai-cheng* (Finance) to be in effect a "process of setting the masses to work level by level."[139] The Chinese maintain that active participation by the masses is a *sine qua non* for effective economic development. Here, contracts are thought of primarily as a means to incite mass action. In this regard the *Chieh-fang jih-pao* has said:

> The second advantage [of the investment-by-contract system] is that in order to carry out the terms of the contract, each department and each unit must launch mass movements, organize the masses, and exploit potentials in all phases including designing, construction, and financing. The masses will be informed of the scope of construction, time limit for the work, requirements about quality, and the amount of investment. In this way, the plans will be made more advanced, more economical, and more reasonable. At the same time, while the masses are given their goals to strive for, everybody will be concerned with the construction project, and the drive for "more, faster, better and more economical" construction will become a self-conscious movement of the vast number of workers.[140]

By giving the masses a broader view of the entire project as well as concrete targets, the regime hoped to make the masses cost-conscious. Once the masses realize that extravagance or waste in any phase of the project may result in a failure to fulfill the aim of economy, it is hoped they will not only be concerned with their own productivity "but also with the costs and expenditures of the entire plan."[141] Furthermore, in view of the concededly high administrative expenses consistently incurred in production and construction prior to this period, it is likely that one of the motives for mass participation in management through vertical contracts is to reduce these nonproductive expenses.

In addition to stirring the masses, the concluding of vertical and horizontal contracts provides a means for coordinating the various units concerned. Chinese articles extolling the system declare that before its application there was "duplication of controls" and "buckpassing" but afterwards there was mutual cooperation under the unified leadership of the Party.[142] As a consequence of this cooperation, everyone is said to work together to guarantee the realization of the targets. Self-inspection, mutual inspection, and mass inspection are said to be carried out to insure the quality of performance under the contracts.[143]

Deterioration of quality and waste due to lack of coordination, bureaucratized administration, and mass apathy have been widespread in Communist production thus far. If the devices herein grouped under the phrase "the institution of contracts" could remedy this, the position of contracts would be secure in the future.

Conclusion

The institution of contracts in Communist China seems to have developed into a carrot-and-stick device to be used by the regime to stimulate the masses.

Such a use in itself, however, is not necessarily inconsistent with the Western notion of contracts so long as a minimal degree of legal freedom exists. In the area of carefully planned contracts in China, that minimal degree of freedom seems not to exist. Abstraction does not appear to be the rule but the exception. It may be, though, that in non-planned transactions and in the less rigorously planned transactions freedom and abstraction are significant.

Furthermore, it is apparent that the term "contracts" in China is used to cover a variety of devices. Most often, as in situations where contracts serve primarily to break down the plan to lower levels and where there is little room for the exercise of freedom by the parties with respect to important terms of the contract, the nature of a contract appears to be assimilated to that of an administrative order. In these situations it is difficult to distinguish between plan and contract. The predominance of the element of authority to the point of the almost complete absence of the element of minimal freedom in such contracts prevents characterizing these devices as contracts in the classical Western sense.

It appears from the above that heretofore the focus of the regime has been on the use rather than the law of contracts. While a legal foundation of contract rules seems necessary for the effective use of contracts, even in a relatively simple role—for example, if contracts are to allocate responsibility, there must be methods for determining what the words of the contract mean in the particular situation and for settling disputes about their meaning—it is clear that in practice the regime has not dwelt on this point. Nevertheless the regime has shown a minimal awareness of the need for laws to provide a framework for contractual relationships.

Chapter III

CONTRACT LAW IN CHINA: A COMPARATIVE ANALYSIS

It should not be difficult to understand that even if many of the devices known as contracts in China are not contracts in the classical Western sense, they may still perform some of the functions that contracts perform in the West. Further, in order to perform those functions it may be necessary that rules—laws—be adopted for purposes of efficiency and administrative convenience.

Many of the rules that appear to be espoused by the regime[1] are surprisingly similar to rules in Western law. On the other hand, while many of the rules are not strange to us, their configuration and interaction often cause different results. This is to be expected since the Chinese context requires that the traditional emphasis of Western contracts on private lawmaking be shifted to an emphasis that has little faith in Adam Smith's "invisible hand," an emphasis that rests on the concept and practice of planning.

The Source

Before discussing the law of contracts in Communist China it is necessary to consider briefly the sources of the information upon which many of the conclusions of this chapter are based. There is no civil code in China. Moreover, in several areas there hardly appears to be a satisfactory method of determining the state of the law regarding a particular subject aside from the source used herein.[2] That source, *Basic Problems in the Civil Law of the People's Republic of China*,[3] hereinafter referred to as *Basic Problems*, is the primary authority in translation and in Chinese as well with respect to contract law in the People's Republic. Because of the heavy reliance placed on this single work, a lengthy discussion of its reliability and applicability is set forth in the notes.[4]

Aside from its didactic passages, *Basic Problems* as a whole resembles a good Restatement of Law in that it not only states what the law is and where it is not settled but also what it should be. With regard to its overall structure, *Basic Problems* is strikingly similar to the Russian Civil Code of 1922[5] and the German Civil Code of 1900.[6] Since the Russian Civil Code is derived in large part from the German Civil Code[7] and since Chinese Communist law appears to have been taken directly from Russian law, this similarity in structure is natural.

In all three systems the structural progression in systematic theory is from the general to the specific. Like the Russian and German Codes, *Basic Problems* begins with a discussion of general principles of the civil law.[8] These principles apply to all civil law relations. From there *Basic Problems* proceeds, after chapters focusing on property relations,[9] to a discussion of the general principles of obligations.[10] These principles apply to all civil law obligations, those arising from contracts as well as those arising from delicts. Following this discussion, *Basic Problems* deals with the general principles applicable to all contractual obligations. Finally, *Basic Problems* discusses individual types of contracts, providing rules applicable to each type.[11]

The Contours of Contractual Obligations

The jurisprudential approach of *Basic Problems* is also reminiscent of German and Russian jurisprudence. An obligation is described as a civil juristic relation between specific juristic or natural persons with "disposing capacity." An obligation arises from the juristic acts of the parties to the obligation.[12]

A juristic act is a kind of juristic fact. Juristic facts are the causes of the arising, altering, and extinction of juristic relations. Juristic facts are made up of incidents and acts. The former category comprises facts that are independent of the will of the parties, such as natural death. The latter category refers to: "the conscious [activities] of citizens or juristic persons, aiming at the rise, alteration, or extinction of certain civil relations."[13]

A contractual obligation is an obligation arising from particular juristic acts of both the prospective obligor and the prospective obligee. It is predicated upon concordance of intentions mutually declared by the parties.[14] In this respect it is distinguished from administrative and delictual obligations, which do not depend upon the agreement of parties.[15]

The distinguishing and characteristic element of contracts is the agreement of the parties. As in France,[16] Germany,[17] and Russia[18] there appears to be no general requirement that there be consideration. It is clear, however, that the lack of this requirement is not as significant as it appears to be on the surface since other functional devices perform the tasks of the institution of consideration in German, French,[19] Russian,[20] and Chinese law.[21]

In defining agreement, *Basic Problems* states that "a contract must be made according to the willingness of the parties."[22]

It is later stated that "willingness" is not equivalent to freedom of contract but is a product of both freedom and discipline. The mythical pure freedom of contract of the common law is rejected as a reflection of bourgeois individualism, as an illusion before the reality of economic coercion that exists in capitalist societies.[23] Thus, very little content is given to the requirement that agreement be on the basis of the "willingness" of the parties. It appears that the requirement of "willingness" in practice often has only metaphysical significance in that it means accepting discipline—that is, contracting as **required** by the regime—which, because it is for the benefit of "the people," cannot be done unwillingly. On the other hand, "willingness," of course, does have a limited content with respect to improper interference with willing (for example, duress).[24]

Formation

Under Chinese law, like Western law, to form a contract there must be an offer and an acceptance in terms of the offer. Since no consideration is required in China to support a binding obligation, the offeror "is bound by his offer within a given period of time."[25] If the offeror during that period of time makes the same offer to a third person, he will be liable to compensate the original offeree for any damages caused by the offeror's failure to keep the offer open.[26]

On its face this approach resembles the general German rule that an offer is binding without consideration for the period stated or for a reasonable time.[27] It is obvious, however, that at least with regard to planned contracts the Chinese cannot accept the German rule insofar as it allows the offeror to determine the binding effect of his offer.[28] This broad freedom of the offeror in Western law is antithetical to the notion of planning and to the administrative obligation, see *infra*, imposed on the parties. Moreover, the general German rule is part of a complex of rules that function together to protect the reliance of the offeree upon the offer and to determine when a binding contract has been concluded.[29]

Basic Problems reflects a singular lack of concern with these latter questions. The question of when an acceptance takes effect is not dealt with in general terms. It may be that this question has simply been omitted in *Basic Problems* and that rules similar to those in the Russian Civil Code normally apply.[30]

However, the inattention to the question of whether an acceptance takes effect on dispatch or receipt also may indicate that the problem rarely arises in China. The problem will only arise when contracts are negotiated by correspondence between parties not in each other's presence. In China the means of communication are still very inadequate. As a result, negotiations by correspondence may be unusual for the relatively unimportant unplanned contracts. Insofar as most unplanned contracts are concluded face-to-face by the parties or agents of the parties, this problem does not exist. On the other hand, contracts concluded among the state-owned or state-private joint enterprises are likely to be governed by plans. When this is true, the parties to the negotiations normally are under an administrative obligation to conclude the contract within a specified period of time.[31] The determination of precisely when the contract is concluded, then, is less important as a party can seek administrative enforcement of the administrative obligation even before the civil obligation has been created. With regard to negotiations between state-private joint enterprises the need for explicit rules to determine when a contract has been concluded is mitigated further by the fact that drafts of all contracts must be approved by the specialized industrial companies before the contract is binding.[32]

A party to negotiations regulated by plans thus does not rely heavily upon the offer itself. This is particularly true where the plan or directives issued by organs charged with enforcing the plan even stipulate who the parties to the contract should be. Moreover, the offeree's reliance on the offer in a planned transaction is different in kind from the reliance of an offeree in a "free" economy in that the latter, at least theoretically, will suffer detriment by his reliance and concomitant foregoing of other opportunities to satisfy his needs. In a fully planned economy, on the other hand, it is arguable that the offeree normally will have few alternatives since most other opportunities already will have been assigned by the plan to third persons.

Notwithstanding the above considerations, it seems the Chinese will have to deal with these problems more carefully. The plan in fact does not cover all transactions within the state sphere and even for those that it does cover it does not stipulate who the parties to the transaction should be. So, in these areas the offeree's reliance on the offer may assume greater importance. Furthermore, the government is trying to stimulate an increased flow of goods throughout the country.[33] The realization of this aim will necessitate more negotiation by correspondence and may therefore require additional consideration of the rules regarding offer and acceptance.

Hence, at least in areas that more closely resemble the situation that exists in the West, rules such as those adopted by Western law may be essential to a smoothly running economy. Moreover, even in areas more strictly regulated by planning, such rules seem to be a valuable means to increase efficiency. The need for predictability is vital to the efficiency of the economy. An efficient economy is not one in which every issue between the parties is unsettled and in which litigation is the normal way to resolve disputes. The parties must be able to determine prospectively with some certainty when they have created binding obligations. In the light of these demands, it seems likely that the government will fix a set of more sophisticated rules to regulate these problems.

Unilateral Contracts

Chinese law, in contrast to German contract law,[34] appears to recognize the concept of a unilateral contract. However, unlike common law thinking about unilateral contracts,[35] the Chinese view appears to be that the party performing the act is the offeror.[36] Thus, the performing party cannot rely upon any offer in performing. He must first perform, that is, make his offer, and only then will the other party be in a position to bind himself.

A common law jurisdiction is probably precluded from this approach by the doctrine of consideration, but, aside from doctrinal aspects, the common law approach seems more in accord with reality. The performing party normally will not enter into performance unless he believes the accepting party desires him to. Thus, the performing party is relying upon the at-least-inferred offer of the receiving party. Notwithstanding the superiority of the common law in this respect, even the common law rule in its pure form has itself proved to be inadequate. It is necessary in normal commercial dealings for the performing party to be able to rely with some certainty on the acceptance of his performance. As a result of these needs, the common law has moved toward developing rules to protect the performing party who has acted in reliance on the offer.[37] It appears that Chinese law will have to develop in similar fashion if its view of unilateral contracts is to be applied generally.

However, the only two contexts in which the notion of unilateral contracts is explicitly said to be relevant in *Basic Problems* are the loan situation and the transport contract.[38] In the former it does not appear that the reliance of the performing party will be great. In the loan situation the performing party is the lender. At least with regard to the common loan of money it does not appear that the lender will suffer much detriment due to his reliance. If the borrower does not accept his offer, the lender can relatively easily make the same offer elsewhere.

In the transport situation, though, it is clear that the offeror—the shipper—may suffer great detriment if he brings his goods to the carrier, thereby making his offer, and the carrier then rejects his offer. Still, the performing party's reliance will be protected in many transport situations by an administrative obligation on the part of the carrier to conclude the contract.[39] Furthermore, it seems that the regime has made a conscious decision to protect carriers,[40] and that it has subjected any interest in the reliance of the shipper to that end. If these are the only factual situations to which the concept of unilateral contracts is applicable, the seeming inadequacy of the Chinese version as a general proposition is irrelevant.

It is not clear why the concept of unilateral contract exists in Chinese contract law. It is not present in German law and its presence in the common law has been criticized.[41] As its relevance in China appears to be limited to specific situations, it may be simply a transitional device for giving confidence to private lenders and favoring carriers.

Elements of A Contract
Every legal system requires that certain terms be included in contracts.

In China all "contracts" must specify the "object" of the contract and the "amount" of the "object" (for example, in pounds, feet, hours, and so forth) contracted for. Price and time of performance are said to be essential elements depending upon the nature of the contract. Apparently, where price is considered essential, the parties should set the price according to the officially listed price or, if none is listed, by negotiation. *Basic Problems,* on the subject of contracts in which provisions regarding price and time are said to be essential, rejects the idea that a reasonable price or time can be implied. The formal reason given for this rejection is that a term the parties have not negotiated should not be inserted into the contract. The actual reason seems to be that implying terms will give the parties "an excuse to evade the problems of price and time, which, according to the nature of the contract, should be solved."[42]

With respect to contracts between socialist organizations the list of required terms is more inclusive.[43] Time is a principal element. The contract should set a fixed period of time for performance; merely providing for performance within a reasonable time is improper. The purpose of this requirement of a fixed period of time for performance is to stimulate the prompt exchange of goods. These contracts also should give the specifications of the goods and the standard of quality. It is not clear, however, whether the failure to include these elements will result in the contract being held incomplete and therefore of no effect. It seems unlikely.

In addition to the aforementioned items, the supply contract must incorporate the terms of the "basic conditions of supply."[44] The supply contract is a rigidly planned contract between socialist organizations that is based on production and distribution plans.[45] The conditions of supply are set jointly on an annual basis normally by the consuming and supplying ministries. They are administratively prescribed terms that cover a variety of aspects of future supply and distribution in detail, such as packaging, methods of payment, and penalties. In the areas covered by these administrative provisions the parties have little or no room for private lawmaking. And, "if the agreement between the two parties is in [substantial] conflict with the [spirit of the] provisions . . ., it is void."[46]

Finally, every contract between socialist entities must include a provision for penalty, to be applied in the event of breach. The amount of the penalty will usually be prescribed by administrative provisions.[47]

These rules relating to the necessary elements of a contract are not generically different from Western law. Whether a contract is to be effective as a means of private lawmaking or as an arm of the government, it is necessary that its terms be sufficiently clear to be understood by the parties and sufficiently definite to allow enforcement by the society. Thus, the demands in

both situations are comparable, and the laws adopted in response to these demands are relatively similar.

Formalities

The formal requirements for particular categories of contracts and the
consequence of a failure to satisfy those requirements are not clearly set forth
in *Basic Problems*. A written form appears to be required at least in contracts
between socialist organizations that deal with "certain important matters of
business which cannot be terminated immediately."[48] The implication is that
a written form is not required for all contracts between these organizations.
This is in accord with the Russian rule that only contracts of this sort above
1,000 rubles have to be in writing.[49] In contrast with the broad Russian rule
that all private contracts over 500 rubles have to be in writing,[50] it is not
clear from *Basic Problems* whether there is such a general statute of frauds in
China. There is a specific requirement, however, that insurance contracts be
in writing,[51] and there may be specific requirements with respect to other
types of contracts. The requirement of writing seems to be imposed for both
evidentiary and supervisory reasons.[52]

The supervisory aspect, though, is dominant where in addition to the requirement of a writing the writing itself must be registered. Sales and leases
of private houses must be registered with appropriate state organs.[53] Marriage
contracts must also be registered.[54] The notarized written form was provided
in the past as well, primarily for purposes of state supervision. That form often was required for processing contracts let to private enterprises. *Basic
Problems* notes that the notarized form has not been uniformly required and
leaves it to the will of the parties whether such a form shall be employed in
the future.[55]

Basic Problems does not discuss the consequences of a failure to comply
with formal requirements. In Russia, depending upon the particular circumstances, failure to comply may result in the application of a rule analogous to
the parol evidence rule,[56] or in avoidance of the contract and subsequent restitution.[57]

With regard to many of these aspects of contracts, the impact of the particular economic context of China on the institution of contracts often is not
apparent. It is clear that the economic context of planning results in severe
limitations on legal freedom of contract. Beyond that it is not clear how the
planning context has affected or will affect rules relating to problems of offer
and acceptance, formality, and so forth. Many of the rules associated with
contracts in Western law seem relevant to the Chinese context. While the particular reasons for the rules may not be the same in the Chinese context, it
appears that these rules are often useful in China at least as a framework

for attaining the uniformity and consistency necessary for an integrated economy.

In the realm of performance of contractual obligations and the remedies provided for breach, however, the impact of the context is obvious. While for the most part the remedies provided are similar to those provided by Western law, the relative weight assigned to each remedy is very different.

Performance and Remedies

Specific Performance

Although modern legal systems underwrite the basic theory that contracts are entered into to be performed as written, the conclusions drawn from this theory vary from system to system. In common law countries, for example, the concept of judicially enforced specific performance developed in relation to objects considered by the court to be unique.[58] It was only in these cases that the normal remedy at law for damages was considered to be inadequate. Further, it was only in these cases that the implications of a credit economy in which most products were fungible (substitutable) in terms of money were irrelevant.[59] Thus, the normal remedy at common law for breach of contract is damages. Only if the obligee can show that damages will be inadequate and if the obligor is still able to perform, may the court order the obligor to specifically perform his obligation.[60] In the alternative to seeking damages or specific performance, when the breach of the obligor is substantial, the obligee at common law may seek rescission of the contract and restitution.[61]

The willingness of the common law courts to award damages for breach, instead of requiring performance in strict compliance with the terms of the contract, reflects the underlying economy in which the market places a value on products and in which like products normally can be obtained for like value. That willingness also reflects the society's focus on contracts primarily as a means of facilitating private transactions and its relative disinterest in enforcing remedies not required by the focus. Allowing the termination of the contract by rescission whenever such termination is fair to the parties in the particular case reflects the same focus and same relative disinterest.[62]

In China, on the other hand, the economy is one of greater scarcity. Therefore, goods are less freely available and money may not be easily convertible into desired goods. In this sense goods contracted for may be considered unique. Moreover, under planning, the allocation of resources and products enhances the character of uniqueness as natural scarcity is intensified by planned scarcity (that is, planned balances under which there is little surplus of supply over demand). Hence, in this context even on the common law approach specific performance seems to be a proper remedy.

In addition, the emphasis on contracts in China primarily as a device for implementing and supplementing the national plan reinforces the demand for specific performance of obligations. Because of planning, the state's interest in the performance of each contract is immeasurably increased. Only by specific performance of obligations can the plan be fulfilled. Substituted performance—damages—will not satisfy the production needs of the obligee. And, as a result of that failing, the obligee himself may be unable to perform his obligations in realization of the plan, thereby adding a link to the chain of failures to achieve the plan that may follow a failure to specifically perform any contract under the plan. Consequently, the normal remedy in China for breach of contract is specific performance. Penalties are provided as an auxiliary remedy to deter the obligor from breaching, to punish him in the event of breach, and often to compensate the obligee. The remedy of damages is also secondarily available both for compensatory and punitive purposes.[63]

In addition to these functional reasons for enforcing strict compliance, it is possible that German theory, which asserts the primacy of specific enforcement as the normal remedy,[64] has made an impact through Soviet law on Chinese Communist law. It is doubtful, however, in light of the pragmatic character of the Communist regime even in the realm of theory,[65] that this impact has been determinative.

The implications of this policy of enforcing specific performance in order to implement the plan reach into the area of precontract disputes as well as into the area of contractual disputes. While it is asserted that normally the civil law obligation between the parties does not arise until the concluding of a contract,[66] the prospective parties to a contract regulated by planning documents frequently are under an administrative obligation to conclude the contract.[67] If the parties were not under such an administrative obligation before the contract was formed, the utility of the remedy of specific performance to plan realization would be undermined. Thus, performance of the precontract obligation will be enforced specifically, and failure to comply will result in administrative penalty.[68]

The ramifications of the policy of specific enforcement are most clearly seen in the context of the supply type contract.[69] The planned nature of that contract affects its formation and enforcement.[70] The supply contract, which is between buying and selling units, may be either a subcontract (also referred to as a local contract) or a direct contract. The distinction between the two may be more a matter of form than of substance, though with regard to Russian law it is thought that by and large the parties to a direct contract have somewhat greater freedom to determine its terms.[71]

Whether a contract is a subcontract or a direct one depends upon the structure for the contract set up by the higher administrative organs. If the supply bureau of the ministry in charge of the supplying enterprise and the general marketing bureau in the ministry in charge of the purchasing enterprise conclude a general contract, the contract concluded by the primary parties in accordance with that general contract is known as a subcontract or local contract. The general contract embodies the principal terms and specifies the parties to the local contract. The latter, then, is concluded to create the civil obligation between the parties and to further define their reciprocal rights and duties. On the other hand, no administrative contract precedes the concluding of a direct contract between the primary parties. In either case, however, it appears that the major terms and the choice of parties to the primary contract is determined by administrative documents—general contracts or others. And, in either case, the primary parties must specifically perform their administrative obligation by concluding the primary contract within a specified time. Should the parties be unable to successfully conclude their contract, "the case is submitted to the higher authorities in charge for arbitration and settlement, and the contract is then concluded on the basis of the decision of the higher authorities."[72] The making of a supply contract not in accordance with the plan will be punished.[73]

Once the contract is formed, it must be executed in accordance with its terms.[74] Since the contract is based on the plan, it cannot be altered or rescinded unless there is a change of plan or unless an administrative unit authorized such action.[75] By and large, the period for performance of the contract is limited to the period of the plan, one year. Requiring or even allowing performance after that period, rather than implementing planning, might interfere with the realization of the plan for the subsequent year.[76] Thus, as in Russia, the remedy of specific performance is ordinarily available only during the plan period.[77] Where, for example, the goods to be supplied are not produced within the year, specific performance of the contract will not be ordered during the following year unless the planning organs have agreed to adjust their plans to the situation. On the other hand, where the goods are properly produced during the year but merely are not delivered as agreed, ordering specific performance will not interfere with the obligor's fulfillment of his production plans for the following year, and it will be ordered.[78] It appears in either case that the obligor will be liable to pay a penalty and/or compensatory damages for his default.

However, in the case where the obligor has performed but his performance is defective,[79] or where performance is delayed to the point that it has lost its value to the obligee,[80] the remedy of specific performance often is not proper even during the plan period. Where the defect does not prevent the purchaser from using the goods, policies directed toward maximum utilization of materials militate against ordering specific performance and remit the purchaser to other remedies. Where the performance is simply defective, the purchase price may be correspondingly reduced by the judging organ.[81] But if the goods are so defective that the purchaser cannot use them, they are to be returned at the seller's expense and presumably the seller will be required specifically to perform.[82]

The instances in which specific performance is not appropriate even during the plan period are those in which ordering such performance would not further the interests of plan realization. As suggested in regard to Russian law: "Where performance has been rendered and the breach consists in a mere delay or in minor deviations from the contract specifications, punishment in the form of monetary sanctions is the obligee's normal remedy . . . It is acceptable to Soviet law because from the point of view of planning nothing else could be more expedient; delay is not curable; insubstantial deviations can be specifically corrected only at the expense of further delay and at a threat of new snags in the plan."[83]

Penalties

Monetary penalties are levied on defaulting parties primarily to enforce strict compliance with the terms of the contract.[84] In addition to the function of deterring deviations from the contract by calling attention to and punishing such deviations, the enforcement of penalties may also serve a quasi-compensatory function.

The penalty is considered a form of security of obligation—that is, a device to secure performance of the obligation—just as is suretyship, mortgage, and retention.[85] A penalty clause is connected to the primary obligation to perform and it is vitiated when the obligation is extinguished.[86] Clauses providing for penalties are required in contracts between socialist organizations. The level of the penalty is normally set by administrative regulations.[87] In this realm of contracts between socialist organizations, where a breach of contract results in breach of the state plan, punishment (monetary or otherwise) is considered appropriate both to vindicate the society's interest and to assure maximum efforts in fulfillment of that interest.

As a general proposition, penalty clauses will be enforced for default regardless of the damage or lack of same to the obligee.[88] However, it appears that the breaching party will only be liable to pay a penalty if he was "at fault."[89] Hence, if *force majeure*, change of plan, or military orders were the cause of the obligor's default, he will not be liable to pay a penalty. The requirement of "fault" seems to be a functional limitation on the duty to pay a penalty for breach, since the main purpose of penalties—to encourage strict compliance—is served only when the party was in a position to avoid deviation.

To fulfill the functions of deterrence, punishment, and signaling deviations, it is necessary that the obligee enforce his rights to a penalty.[90] Although *Basic Problems* does not suggest that the obligee is under a duty to seek redress even where he has not in fact suffered damage, it is likely that such a duty must be inferred. If no such duty exists and if the obligee has not suffered damages in fact and therefore has no affirmative reason to bring the default of the obligor to the attention of the administration, defaults will go undetected and the policy of strict compliance with contracts will be undermined. In Hungary, for example, it appears that even when such a duty exists the obligee is prone to avoid enforcing his rights where he has suffered no damage, in expectation that his obligor may one day be his obligee, who can then return the favor by similar forbearance.[91]

On the other hand, unrelenting assertion of rights to penalties for purely formal breaches will result in unnecessary litigation and administrative paralysis. In this light it seems necessary in practice to allow the obligee some degree of discretion in deciding when to raise the issue. By doing so, of course, the administration runs the risk of diluting the policy of contract discipline.

At least with regard to unplanned contracts, where penalty clauses are optional, the court can reduce the amount of the penalty in the light of all the circumstances, including the obligor's financial position and the effect on his productive capacity of imposing the penalty.[92] The court, however, cannot raise the penalty above the contracted amount.[93] It is probable that the administrative organ judging a planned contract dispute is given the same latitude, even though the penalties in that case have been set by administrative decrees. Since the main aim of the penalty system is to contribute to the realization of the plans, it would be anomalous to allow penalties to be levied in such a way as to interfere with the particular obligor's capacity to produce, thereby interfering with realization of the plans.

Penalties also serve a compensatory function. This dual use of penalties was adopted from the Soviet experience.[94] Where the obligee suffers damages below the amount of the penalty, the penalty itself is compensatory to the extent of the damages, and "the court may properly reduce the . . . [obligor's] liability to pay . . . compensation."[95] In that case the penalty, unless reduced by the court, in effect provides an assured minimum recovery for the obligee, regardless of the extent of actual damage.[96] In the event the penalty amount does not cover the damages suffered, the obligee can recover compensation damages to make up the difference.[97] Thus, only when no actual damages are suffered does the penalty work purely in a noncompensatory manner. In view of the above-suggested natural disinclination on the part of parties to enforce a penalty clause where no actual damage has been sustained, the significance of the purely penal aspect of penalties in practice may be slight.

The common law disaffection with penalty clauses reflects to a great degree the relative unimportance attached to the remedy of specific performance in the usual transaction. The general common law rule is that a liquidated damages clause will not be considered a penalty clause and therefore will be enforced only when both the damages are such that they would be difficult for the court to compute and the amount set as liquidated damages is within the range that prospectively would have been regarded as reasonable compensation for any breach to which the clause might apply.[98] This rule reflects the beliefs that damages are normally easy to calculate in a market economy, that damages are essentially compensatory and should not perform other functions, and ultimately that specific performance of obligations is not so important as to justify the imposition of penalty clauses. Without the compulsion supplied by the context of a planned and scarce economy, the common law adheres to the principle that compensatory damages generally are the adequate and proper remedy.

Damages

The nature and functions of the remedy of compensatory damages in China are not unlike those of the remedy of penalty. Although damages will not be awarded unless the obligee has suffered injury and the extent of the award will be determined by the extent of the injury "caused" by the obligor's default,[99] nevertheless, *Basic Problems* treats both penalties and damages as "legal punishment" for non-performance.[100] It justifies this characterization on the ground that both remedies impose new obligations on the obligor.[101] Of course, this explanation applies as well to all remedies for breach of a delictual obligation,[102] and it implies that all compensation is punitive at least in part. While this approach is not dominant in common law thinking about contracts, it seems that in both systems there is a punitive and deterrent aspect to compensatory damages. The greater emphasis in China on the punitive aspect is understandable given the higher degree of affirmative and direct state involvement in the primary relationship between the parties and the Communist emphasis on the educational value of law.[103]

The remedies of penalties and damages will normally be available as supplementary to the primary remedy of specific performance. Only when specific performance is impossible or will not contribute to the realization of the plan will the secondary remedies be appropriate as alternatives to specific performance. All three remedies act as devices for carrying out administrative supervision of contracts. The assertion of remedial rights signals and defines the scope of default. And, the award of penalties or damages adversely affects the financial plan of the obligor, thereby flagging the breach of the obligor.

Under the system of business accountability the books of the obligor will show his default, thus bringing it to the attention of administrative authorities at the end of the fiscal period as well as at the time the contract dispute is litigated. Insofar as both remedies are compensatory, they also impart that the failure on the part of the obligee to realize his production plans is not attributable to him but to the obligor. This will be apparent under the system of business accountability by the obligee's simultaneous fulfilling of his financial plan (as the award he receives will compensate for any loss of planned profit) and his failing to fulfill his production plans (as the obligee did not receive the necessary goods from the obligor).

Under this system the need for the remedy of compensatory damages is manifest. If the system is to operate properly, the allocation of responsibility must be reflected in the accounts of the concerned enterprises. Hence, where penalties do not suffice to cover the amount of damages suffered, compensatory damages are needed to reflect the lack of responsibility of the obligee in failing to fulfill his plans. Furthermore, compensation is often necessary to insure the ability of the obligee to produce in the future.

The normal measure of such damages seems to be the same as common law expectancy damages.[104] Although the Russians and the Chinese are predisposed ideologically against the recovery of future profits, both have accepted the general rule that compensation will be awarded for "the positive loss of property and the loss of obtainable benefits. The former refers to the reduction of existing property, and the latter to the loss of benefits which could otherwise be obtained under normal circumstances."[105]

This result seems necessary in the context of the business-accountability system, in which accounting units are judged to a degree by the planned profit their operations show. Where profit constitutes one standard for measuring performance, expected profit cannot be excluded from the measure of damages.

In the realm of nonplanned contracts the direct interest of the state is less. But insofar as the business-accountability system is relevant, it appears the expectancy measure is still appropriate. However, the common law reasons for adhering to that measure probably are less relevant. The common law reasons[106]—to encourage reliance on contracts, to compensate the obligee for the other like opportunities he abandoned, and to eliminate the problem of alternatively proving the loss in terms of missed opportunities due to that detrimental reliance—could be satisfied by granting reliance damages where, as in China, the obligee may have few other opportunities and therefore his damages tend to be essentially out-of-pocket expenses. Nevertheless, because it is an oversimplification to say the obligee has no other opportunities or possibly because of the at-least-tangential relevance of the business-accountability system to nonplanned contracts, the measure even in nonplanned contracts appears to be expectancy.[107]

The extent of the recovery for damages under the expectancy measure is limited by: (1) the legal rule as to cause and effect; (2) specific regulations regarding certain contracts; and (3) the economic status of the particular parties.

(1) *Basic Problems* states that the obligor is liable for compensation only when the "certain damage [to the obligee] is the inevitable consequence of non-performance."[108] That this rule does not solve problems but merely creates them is apparent. It is not enough that the obligor's nonperformance is a necessary cause in fact of the subsequent injury.[109] Therefore, a showing of mere retrospective inevitability in fact is not conclusive on liability. Some principle, then, must operate to limit liability. Notwithstanding the alleged Russian rejection of the *Hadley v. Baxendale* (1854), 9 Exch. 341, rule of foreseeability as the standard for limiting damages recoverable,[110] and notwithstanding the influence of Russian law on Chinese law, it appears that in essence the foreseeability rule is the applicable standard. By analogy to tort law in China, an analogy that is expressly suggested by *Basic Problems* in the discussion of damages due to the breach of any obligation, it appears the rule of foreseeability limits liability for breach of contract.[111]

Some cut-off line is obviously necessary if the extent of liability for compensation is not to reach to remote damage. In the context of a planned economy, where at least theoretically many losses will be foreseeable, the foreseeability rule seems appropriate.

(2) Specific regulations also may arbitrarily limit the liability of the obligor in certain contracts. For example, in order to protect railroads from the risk of overly burdensome liability, the railroad is "not liable to pay compensation for the loss of obtainable profits."[112]

(3) In addition, in determining the extent of recovery the court takes the relative financial positions of the parties into account. If the obligor can afford to pay the full amount of damage, he will be so liable. If, on the other hand, imposing full liability on the obligor will work undue hardship on his dependents (in the case of a private contract), or will unduly interfere with his productive capacity, liability may be restricted.[113]

Impossibility

Damages, like penalties, are not recoverable unless the obligor is "at fault." It seems clear that by "fault" the Chinese do not mean negligence in the common law tort sense. Rather they appear to mean "responsibility," which appears to be presumed unless nonperformance is shown to be due to *force majeure*, change of plan, or military orders.[114] Thus, no affirmative showing of negligence would be necessary for the obligee to prove contractual "fault." It is not certain whether *force majeure* extends to the breach of an obligation to the obligor by a third person, which breach itself was the cause of the obligor's default. While such an event is not within the normal definition of *force majeure*, and while the common law normally would not recognize the third person's breach as equivalent to impossibility,[115] it would not be unreasonable in light of the scarcity of goods in China to consider such a breach as tantamount to impossibility.[116]

The ramifications of the concept of impossibility in China and Russia are unclear. In Russian law it apparently includes legal impossibility as well as factual impossibility. Thus, where an export license was denied, performance was held to be excused.[117] The Chinese view at least extends to this holding. Moreover, the language in *Basic Problems* used to justify this deviation from strict contract discipline is broad enough to include even the breach of a third person. Failure to perform is said to be excused when "the non-performance of the contract is not due to the fault of the parties, but due to the restriction of objective conditions."[118] If the Chinese do recognize third-person default as a basis for impossibility, thereby bringing tort and contractual concepts of liability closer together, the ultimate responsibility of the contract obligor is significantly limited. As his risks are decreased, he is less in the position of an independent party to a contract as understood at common law, and the classical Western conception of the contract is further impaired.

Avoidance

The rules relating to voidable and void juristic acts are familiar ones. Fraud, "great misunderstanding" by one party caused by the fault of the other party, physical duress, and improper official orders provide a basis for avoidance of the obligation on grounds of the avoiding party's lack of "willingness" in consenting.[119] In addition, as in German law,[120] where a contract is obviously unfair and the oppressed party is in an unequal bargaining position, the contract is voidable.[121] With regard to contracts that are simply grossly unfair, but where there is no disparity of bargaining positions, at least the unfair portion will be annulled as to future performance.[122]

Acts that are done by juristic persons lacking disposing capacity and those that are merely sham are void.[123] In addition, acts that violate "the constitution, the law, the state plan, public order, or social ethics" or acts whose purpose "is to damage the public interest" are void.[124]

The effect of avoidance in China resembles its effect in the Soviet Union.[125] The innocent party, for example, the defrauded party, is entitled to restitution and to compensation for damages caused by the fault of the other party. The guilty party, however, is not entitled to restitution, and his property is forfeit to the state. If the case is a serious one, it is suggested, the state may seize the property the guilty person should have transferred as well as the property he actually transferred.[126]

Where the contract is illegal, the provisions regarding state seizure are also generally relevant. But if the illegality is only slight, the goods involved in the contract are subject to compulsory purchase by the state at a price set by the state instead of outright forfeiture.[127]

It is interesting to note that the Chinese definition of illegal acts and the provision for forfeiture do not follow the literally more encompassing provisions of Articles 30 and 147 of the Soviet Civil Code.[128] Thus, while under Article 30, which provides that "a transaction directed to the obvious prejudice of the state . . . shall be invalid . . . ,"[129] the subjective intention of the parties is irrelevant, under Chinese law it appears that only acts whose purpose is to damage the public interest will be invalid.[130] Similarly, while under a literal reading of Article 147 the intention of the parties is irrelevant to the issue of forfeiture, the Chinese seem to ground forfeiture on subjective guilt. This is in accord with Article 147 as it subsequently has been interpreted by the Soviet courts.[131] The narrowing of the scope of invalidity and forfeiture in some degree enhances the integrity of contracts and the freedom to conclude them. Still, that freedom is limited.

While the scope of direct government intervention by means of avoidance is broader than in Western societies, the fundamental reasons for avoidance—most of which appear to be based on broad notions of justice—are comparable. The fact that the Chinese regime plays a more affirmative role in this respect again only reflects its more intense concern with all aspects of the economy and its decision that government action is necessary to economic development.

Conclusion
The institution in the People's Republic of China that is commonly referred to as contracts is not monolithic. The term "contracts" is applied to a spectrum of devices ranging from those that leave no room for significant private lawmaking and are therefore little more than administrative obligations, to those that allow more individual freedom and therefore more closely

approximate contracts in the Western sense. In the former category are contracts most vital to the regime—the contracts that directly implement the national plans relating to important products. For obligations in this category the term contracts in its classical sense seems to be a misnomer. The other category comprises obligations in which the mandatory planned element is less encompassing, in which the parties themselves can and do stipulate some provisions. While the degree of involvement of the Chinese government still appears to be greater in this category than that of other governments in most contracts in the West, it should be remembered that "freedom of contract" in its pure sense is a fiction in the West as well. By and large, contract terms such as wages, hours, and so forth are regulated by all these societies. Moreover, with regard to public utilities, for example, the degree of government regulation of contracts in the West is very high, and insofar as these obligations are considered contractual in the West, it seems the term is not inappropriately applied to the category in China under discussion.

However, it is important to note that this approach seizes on a class of obligations in the West whose nature is *least* indicative of the nature traditionally ascribed to contracts and uses this classification as a basis for the argument that the analogous category in China also can be called contracts. In other words, this approach takes a class of obligations in the West that is at one end of the spectrum called contracts as it exists in the West and argues by analogy that a category at the opposite end of the spectrum as it exists in China therefore can be called a contract. Although the matter may be one essentially of semantics, it may be more useful to consider both as outside of the realm of contracts. Under this view, then, the absence of the element of minimal freedom in China may preclude classifying such obligations as truly contractual.

Nevertheless, because the regime wants this institution to perform functions of facilitation of economic development, stimulation of the individual, and allocation of responsibility comparable to those in the West, it has repeatedly encouraged the extension of contracts and eulogized the importance of individual initiative. Aside from the use of government processing contracts to transform the private sector of the economy during the early years of the regime, the emphasis of the regime has been on implementing planning in various ways through contracts. Recognizing the need in plan execution for mass enthusiasm and personal initiative, the regime uses contracts to fulfill that need. Yet it is unwilling to allow mass enthusiasm and personal initiative to impair basic elements of the plan, handed down from above. Contracts, embodying such inherently contradictory aims, merely reflect rather than resolve the tension between freedom and authority. Often described by the regime in terms of freedom, they in fact appear to be formed and to operate primarily within the framework of rigid authority. Therefore, it appears that to the degree the aforementioned functions can be carried out by purely administrative devices the institution of planned contracts in China can perform them. But, on balance, it is doubtful whether contracts in China can do more. In this sense, contracts may be indistinguishable from plans.

Rules of law regulating contracts, like those regulating plans, can help contracts perform these functions. Rules relating to such aspects as formalities, required terms, offer and acceptance, and so forth, may increase the efficiency of the institution's operation, much as they do in the West. Whether as an administrative device or as a method of private lawmaking, it is imperative that the contract obligation be definite and clear enough to be understood and enforced. These rules provide for that clarity and definition.

Furthermore, in the context of a planned and scarce economy, remedial rules in particular serve to promote the purpose of contracts in implementing planning. Thus, as a general rule, specific performance of contracts will be enforced whenever such enforcement is in the best interest of plan realization. Penalties are provided primarily to encourage strict compliance with contracts, and they will be assessed for breach thereof. Compensatory damages also are provided to round out the remedial rights by compensating the obligee for injuries beyond the amount provided by penalty. This configuration of remedial rights is in marked contrast to their relative arrangement at the common law, where damages are the normal remedy, specific performance an "extraordinary remedy," and penalty clauses, as such, unenforceable.

The contrast reflects the different needs of the two kinds of societies and the differing responses made at the level of law to those needs. In China, goods, because of natural and planned scarcity, assume a kind of uniqueness that is in contrast with the availability of most goods in the West. Therefore, the need in China is greater for requiring specific performance of obligations, which for all practical purposes may not be easily available elsewhere. This need is further increased by the demand of planning that obligations be properly performed so as to implement the plan. The society, as a party directly interested in specific performance of contracts, protects its interests through the complex of remedies.

Chapter IV

CONTRACTS IN CHINESE AGRICULTURE
A CASE STUDY OF CONTRACTS IN PRACTICE, 1949–1963

For a Western-trained lawyer considering contracts in action within the context of a planned economy, three caveats are in order. First, plans inevitably involve consultation, which easily shades over into negotiation at the stages of formulation and implementation (the latter often requiring reformulation). In this sense, plans are not wholly antithetical to the concept of agreement between concerned parties. Second, subject to that qualification, a primary function of contracts in a planned economy is to implement the planners' preferences. Consequently, it is to be expected that the degree of private autonomy associated with contracts in the West will be restricted. Finally, to repeat, in Western societies also the parties are limited, to a lesser degree to be sure, in the terms they may incorporate in a contract; for example, minimum wage laws set limits on the possible wage provisions in contracts. Thus, in the theoretical spectrum from wholly unrestrained, enforceable agreements to pure commands, both plans and contracts in practice fall somewhere in the middle in all societies.

In China, plans and contracts are two parts of the same process of defining future expectations and behavior.[1] This process cannot be subsumed accurately under an unrefined notion of command. In many respects it is more akin to the familiar circumstance of negotiations between parties in unequal bargaining positions. The impact of the unequal bargaining position varies with the particular context. Generally, the assertion of its superior bargaining position by the government, that is, government influence, is greatest when the contract is concluded with a unit the state wishes to restrict, transform, or affirmatively support and when the object of the contract is a product that is important to the state. Concluding contracts is one of the basic methods the state employs to organize and coordinate different units within the economy. Contracts in China, therefore, for the most part appear to be less agreements reached through private negotiations by the parties than predetermined bargains that one or both of the parties are induced to accept.

Insofar as contracts are dominated by the "command" element, understood always in terms of the state's superior bargaining position, it is arguable that there is no reason to call such agreements contracts at all. Yet, to the extent that the "command" element is less dominant than in plans, it is still useful to distinguish contracts from plans. Moreover, since even the most rigidly planned contracts involve civil remedies not ordinarily attaching to plans, there is some reason not to lump plans and contracts together.

My discussion of agricultural contracts in China begins with an analysis of their role and development from 1949 through 1957 and then presents a number of constants in contracts during the entire period from 1949 through 1963. That is followed by a detailed look at institutional developments in China's rural economy after 1958. These developments illustrate how contracts are integrated into rural marketing institutions. Finally, this chapter deals with the severe contradictions in Chinese contracts, drawing for comparison upon recent trends in Soviet contracts, and examining several relevant theses of Professor Franz Schurmann.[2]

Development and Role of Contracts in Agriculture, 1949–1957

The contract system (*ho-t'ung chih-tu*) in China has been used consistently in agriculture as an instrument of state policy. The system has been praised profusely during all periods since liberation.[3] While the emphasis from area to area and period to period shows variation in the use of contracts, it is the system's singleness of purpose rather than its variations that is most striking.

Contracts appear to have played three complementary roles in rural China since 1949: (1) before 1957 in the collectivization of agriculture; (2) in implementing state plans by concretely bringing targets down to production units, thereby linking the unit's production plan to state plans; and (3) in supplementing state plans by guiding and organizing production and marketing of products not included in the plans.

Supply and Marketing Cooperatives

The supply and marketing cooperatives (SMCs) have been the vital institution employing contracts in agriculture. Although technically the supply and marketing cooperatives are "mass organizations" (*ch'un-chung-hsing t'uan-t'i*) managed on the principle of democratic centralism, in fact these cooperatives like other such organizations have functioned primarily as appendages of the state. The basic-level SMCs in rural areas are the omnicompetent, unofficial agents of state economic organs at the local level, often below the penetration of the official state hierarchy.[4]

The SMCs act as middlemen between the state and rural production units and, on behalf of the state, between rural production units themselves (for example, between agricultural and handicraft production units). Most of their purchasing work is for the state.[5] All or a very large portion of commodities such as grain, cotton, and other important raw materials for light industry and export are purchased through the SMCs. The SMC hierarchy, in performing its vital functions for the regime, has relied heavily on contracts.

Kinds of Contracts

Chinese sources reveal that there are two major forms of contracts in agriculture, the advance purchase contract and the combined contract.[6] Despite the fact that one form is referred to as an *advance* purchase contract and the other is not, both in fact involve the purchase in advance of crops or other commodities usually not in existence at the time the contract is signed. Since their inceptions, and particularly in the early 1950s when they were developing, there often has been a confusion of terminology and content between the two forms. There also have been regional differences. At times the distinction between the two forms seems to have ceased to exist.[7] Still, in theory at least the two forms are said to be different.

Theoretically, the purchasing side in an advance purchase contract makes a monetary advance payment when the contract is signed. This advance is accompanied by a certificate issued by the purchasing side—usually an SMC—which entitles the producer to priority in buying materials whose supply is limited. The producer is not obligated to purchase these other materials but may exercise his preferential right within a fixed period if he wishes.[8] Thus, the directives of the cabinet relating to the state advance purchase of cotton for 1955[9] and 1956[10] specify that on the average a monetary advance of 10 per cent should be made and that the right to a priority supply of grain and cotton cloth, in addition to the amount allocated under unified sales,[11] should be in direct proportion to the amount of raw cotton sold to the state.

The distinction between the priority right of the producer to purchase other materials and his obligation to do so is stressed because it is this assertedly nonobligatory aspect of the advance purchase contract that is said to distinguish it conceptually from the combined contract. In the latter both sides are mutually obligated to buy and sell, with the SMC, for example, purchasing agricultural commodities from the production unit and in return supplying it in advance with materials needed for production or for the livelihood of its members.[12]

It is hard to determine whether this theoretical distinction between the advance purchase contract and the combined contract exists in practice. The evidence is conflicting.[13] Foreign scholars, however, are not the only ones confused. A lecturer in civil law at a university on the mainland apparently concluded, erroneously we are told, that "it is only the name that is different."[14]

But, at another level, there is a significant difference between the two types of contracts. The difference is not formal but functional, based on the organizational needs of the state for managing production and marketing. The difference appears to lie in the nature of the products involved in each form and in the concomitant degree of centralized planning attaching to each.

For the most part, advance purchase contracts have dealt with crops that are most vital to the state, especially those which since 1953 and 1954 have been subject to unified purchase and sales by the state.[15] These crops, like cotton, grain, and oil-bearing plants, have been consistently under strict state control and planning. Since the implementation of the unified purchase and sales system, the most vital of these commodities cannot legally be purchased or sold except through state commerce.[16] The production and trade of these products is exclusively managed by state organs and state-operated special trading companies, often through the SMCs that act as their purchasing agents in contracts with production units.

Advance purchase contracts appear to be miniature replicas of state plans for the products involved, fixing acreage to be planted by the production unit, production goals for each *mou*, total production, and the portion of total production to be sold to the state.[17] Like all agricultural contracts in China, they are intended both to reflect the actual capacity of the production unit and to stimulate increased production and marketing. In addition, a related aim of these contracts, again like the combined contract, has been to encourage collectivization. Thus, although as late as December 1955 it was still considered proper to conclude advance purchase contracts with individual peasants, more advances were to be made to mutual aid teams and agricultural producers' cooperatives than to individual peasants.[18]

Since the advance purchase contract involves crops most vital to the regime, it can be thought of as more important to the national economy than the combined contract. The rule appears to be that when advance purchase contract obligations conflict with those under a combined contract, the former must be performed first.[19] But, in another sense, the advance purchase contract may make less of a unique contribution to the organization of production and marketing than the combined contract. The advance purchase contract deals with a limited number of crops, many of which are relatively amenable to direct control by planning units, since these crops often are grown in large amounts in rather concentrated areas. The combined contract, on the other hand, deals with products which are relatively less amenable to direct centralized planning, products produced in a more dispersed fashion and on a smaller scale. Without combined contracts, it would be more difficult to exercise even minimal control over such products.

In general, since the establishment of unified sales and purchases in 1953 and 1954, the combined contract has covered agricultural and sideline products, such as green vegetables and chickens, that normally are not included in higher-level state planning. The "combining" that takes place through this contract, while inherent in any bilateral exchange contract, has been enshrined as unique by the government. Ideologically, the Chinese look favorably on the combined contract as a step toward the direct exchange of goods without the use of currency.[20]

The contents of the model combined contract as announced in 1952 clearly reflect its underlying goal of stimulating cooperativization and organizing agricultural production and sales.[21] The degree of generality and the scope of this contract make it apparent that the model contract is less a contract in the Western sense than an institutional framework to be established in rural China, under which the SMC system can extend and consolidate its control of the rural market, thereby ultimately displacing rural private commerce.

The development of the combined contract, like that of the advance purchase contract, has been uneven.[22] With the organizing of 96 per cent of all peasants into agricultural producers' cooperatives (APCs) by the end of 1956,[23] it was said that the resulting increased capacity of SMCs and state organs to organize production created favorable conditions for the signing and performance of contracts.[24] Yet the reopening of the free market in mid-1956 appears to have offset potential gains to contract performance from co-operativization.[25] Peasants sold their products on the free markets, ignoring their contracts.[26] It was difficult to control the scope of the free markets, which were intended to be limited primarily to retail trade in small-scale products of the private plots and APCs. As a consequence, the pattern of advance purchase and combined contracts supplemented by free markets did not work. Recognizing this fact, the cabinet in August 1957 explicitly restricted the free market.[27]

Constants, 1949–1963

The nature of controls, the methods for signing contracts, the organizational structure for assigning contract tasks, the channels for resolving disputes, and the problems that cropped up all seem to have changed very little during the entire period. These constants are a function of the instrumental use made of contracts by the state and of the severe contradictions inherent in contracts in Communist China.

The Parties and their Bargain

The parties to agricultural contracts for the most part have been agricultural production units on one side and state organs or their delegees, the SMCs and special trading corporations, on the other. Contracts normally have been concluded in direct meetings or at conferences where the state dominates.

Where the state or its agent is a party, the ideal for concluding contracts is said to involve investigation by the state side of the actual conditions of production, supply, and marketing, discussion within the production unit of its needs and capacities, followed by a contract based on the interests of both sides. In fact, unrealistic contract tasks more favorable to the state than to the other side often have been delegated from above and were not subject to mass discussion. Published reports and individual interviews confirm that the contract task frequently has been simply announced by Party or state cadres to the leaders of the production unit who, in turn, announced it to the members of the unit.[28] In this context, political education and *shuo-fu* (persuasion), which in practice often means convincing the production unit to accept for the most part what the state offers, have been considered a vital part of contract work.[29] Moreover, the scope of the bargain is severely limited. On the all-important question of price, for example, it seems that the government-set price (*p'ai-chia*) is controlling for all first- and second-category items and even for important third-category items. Where there is no state-set price, the parties are exhorted to use state prices for similar items as a guide.[30] Interviews with former cadres suggest that for many products the scope of bargaining about price may be limited to negotiating the time and place of delivery and other such items which indirectly affect price.

The Integrity of the Contract

The contract as signed is very vulnerable to outside interference. Since contracts serve the planned economy, the integrity of a contract will be sacrificed, when necessary, to the needs of the national economy.[31] If plans are changed, related contracts are expected to be changed accordingly.[32] Moreover, in order to take advantage of favorable conditions like a bumper harvest, the contract task appears to be "elastic"; it may be increased. In fact, revisions of contracts, both upward and downward, the latter in the case of natural disasters, appear to be very common.[33] Though, in principle, such revisions require mutual agreement, the ultimate decision seems to lie with the state.

Vertical and Horizontal Integration

Basic-level contracts in China are expected to be horizontally and vertically integrated with the rest of the economy. Within limits, plans can be used to effect both types of integration. Horizontal integration is also provided by the local unit on the state side that often organizes local sources of supply and credit for the production unit. These links frequently have been formalized through connecting, processing, or order contracts made with handicraft units to supply agricultural production units, and through connecting loan contracts. They also have been formalized through three- and four-cornered contracts in which credit cooperatives and handicraft units are all parties to the primary contract.[34]

Vertical integration is provided also by the hierarchical chains of command in various institutions like the Party, government, SMCs and the materials-exchange conferences. Lower-level units in these hierarchies are expected to carry on their business in accordance with plans, internal directives, and general contracts. The general contracts, signed at higher levels, lay down norms to be fulfilled through concluding more concrete contracts at lower levels.

Vertical control is manifest as well in the settlement of disputes between parties. Settlement is effected, interviews indicate, not through recourse to the courts, which deal almost exclusively with a limited range of criminal matters and private civil disputes,[35] but through decision of concerned government (and Party) administrative units. In general, these administrative decisions aim at insuring plan fulfillment and implementing official policies. Little information has been found specifically regarding the settlement of contract disputes in agriculture. One must refer therefore to industrial contract disputes.

The theoretical pattern for resolving contract disputes appears to have been set at least as early as September 1950, in connection with contracts between state-operated enterprises, state organs, and cooperatives. It was then provided that if both parties were in the same large administrative area and one party failed to perform, the aggrieved party could apply to the local government Finance and Economics Committee for disposition; or, if the parties were not in the same large administrative area, the aggrieved party could apply to the cabinet's Finance and Economics Committee for disposition. It was further provided that if the disposition was "without effect," then the aggrieved party could bring a complaint in the People's Court.[36]

While recourse to the courts is permitted in theory,[37] in practice it appears that the parties have rarely appealed to them. The pressures to conform to administrative dispositions on the whole appear to be so great as to discourage going to court. If both parties are under the direction of the same state unit (for example, a light-industry bureau in a city), administrative resolution of the dispute is a relatively simple matter, to be handled within that unit. If the parties are under different state units or in different areas, however, the dispute apparently is resolved by a joint meeting between the units or by a higher-level unit, such as the provincial Staff Office for Industry and Communications, which has jurisdiction over the lower-level state units involved. Interviews indicate that the ultimate administrative decision frequently has been shaped by Party men within the administrative unit concerned.

A refugee report of a dispute between a cotton textile factory and a dye factory, which was handled by the Tientsin Textile Industry Bureau in 1959, suggests that this process can produce reasonable decisions. In this case, D dye factory contracted with C cotton textile factory for the manufacture of X amount of cloth. C failed to perform and, as both parties therefore failed to complete their plans, the bureau sought to assign responsibility. D blamed his plan failure on C's breach of contract. C, in turn, said his failure to perform was ultimately due to a failure in the state supply of cotton. C refused to compensate or pay D penalties. The bureau held a conference at which the parties presented their sides of the dispute. Then it set up an investigation team to look into the facts. The report of the investigation team substantially supported C's story. On the basis of that report, the bureau held that C need not compensate D since the breach of contract ultimately was not his responsibility.[38] It also held that the relevant plans should be reduced to coincide with actual deliveries made by C to D.

This informal, pragmatic approach to resolving contract disputes suggests that at present, within the Chinese context, the need for a complex body of contract law is minimal. Contractual rules, such as those governing when an offer or acceptance takes effect, become largely superfluous when contracts are hammered out at state-supervised conferences attended by all the parties and when the contract is not the product of a relatively freely-negotiated bargain between equals. Even rules relating to compensation and penalties, which one might think were minimally necessary to carry out the highly-praised system of business accountability (*ching-chi ho-suan chih-tu*), rarely seem to have been systematically enforced.[39]

Primary reliance appears to have been placed instead on administrative rather than contractual rules and upon the complex and comprehensive administrative system, involving vertical and horizontal integration, in which contracts are embedded. Unless the element of private autonomy in contracts is increased, formal contract rules appear to be of only minor importance in China.

The Effectiveness of Integration

By emphasizing vertical and horizontal controls, I do not mean to suggest that these controls have been highly effective. Notwithstanding all the attempts at vertical control, it is very difficult to say to what degree such control has been successful in systematically directing basic-level contracting. The degree of effectiveness has undoubtedly varied from period to period and place to place. Yet, even when it was most effective, it appears to have been unable to solve problems that have plagued the contract system since its inception.

The peasants in 1961, for example, seem to have had the same fears and suspicions about signing contracts that they had in 1950.[40] Frequently the peasants would rather sell their produce on the free market at higher prices than sell to the state.[41] In this respect they appear to have been impervious to political education.[42]

The rural cadres themselves repeatedly have done their work in a mechanical, careless fashion, often simply concluding contracts as a formality, without adequate preparation and supervision to guarantee performance.[43] Organization of local sources of supply frequently has been slipshod, and materials supplied by the SMCs often have not been suitable to local needs. Overaccumulation and shortages have been commonplace. Cadres are still said to be inexperienced and to lack understanding of the solemnity of contracts.

Many of these problems appear to be inherent in the Chinese system of contracts. From the point of view of the agricultural production unit which is a party to a contract, it hardly seems to matter whether the contract is inspired by higher-level planning (as in advance purchase contracts) or, in an apparently more diffuse manner, by the SMC system (as in combined contracts). In either case, interviews indicate that the production unit, in an unequal bargaining position, accepts a largely predetermined bargain. When the contract is more like a little plan—within limits, imposed from above—than like a negotiated exchange between equals, contracts suffer from the same defects as plans.

Thus, although one of the main purposes of contracts is to give the masses a sense of participation in order to stimulate production, it is doubtful whether this goal has been attained. In 1955, for example, when many members of mutual aid teams and agricultural producers' cooperatives withdrew in reaction against cooperativization, it was said that some of them felt the contract was not something they themselves had signed and that its performance therefore was not their concern.[44] Contract performance, in fact, on the whole seems to have been quite uneven.[45]

The Peasants' Interests

At least in principle, however, the system of combined and advance purchase contracts is not without its attractions to the traditionally hard-pressed peasant. All things considered, the poor peasants undoubtedly are better off with this system than they were without it. It promises them both an assured market for their goods and a supply of needed materials for production and livelihood that can be purchased on credit. Moreover, at least until 1954, price incentives were incorporated into state purchase contracts, even with regard to major agricultural products.[46] However, when put in perspective, these benefits are qualified. In practice the contract system's primary purpose is to bring major and secondary agricultural and sideline products into state commerce and to market available manufactured products in rural areas. The

peasants' immediate interests appear to have been given secondary consideration whenever the regime felt that those interests conflicted with the system's
primary purpose.

Post-1957: Institutional Developments

The changes in China's rural economy that occurred during the Great
Leap and communization in 1958 are well known. Suffice it to say that the
elimination of private plots, the end of rural markets, and the extreme antip-
athy to commerce during 1958 reflected a total rejection of recently consoli-
dated patterns.[47]

By the spring of 1959 the government, or a group within it, appears to
have realized that new measures might be necessary to improve the flagging
rural economy. In a series of interrelated announcements from February to
September of that year, the government set up a framework for reviving the
pattern in commerce it had rejected in 1958. And, interestingly, it strengthened the institutional structure for the combined contract, presumably to protect it from the rural market (*chi-shih*) whose revival it soon was to announce.

Institutional Developments in 1959

First, in order to make perfectly clear which products were of greatest
concern to the government and which therefore would be most strictly controlled, the cabinet approved a joint report of various ministries dividing merchandise into three categories.[48] In general the first category corresponds to
products that were previously subject to unified sales and purchases. The
second corresponds to other important products that normally were within
the range of state planning. And the third category comprises the vast residue of items that have rarely been planned at either the central or provincial
levels. Although the total number of products in the first two categories
(331) is small in comparison with the myriad of products in category three,
the products of the first two categories are dominant both in terms of importance to the national economy and total value.[49]

Second, on March 18, 1959, the cabinet approved a report from the
Ministry of Commerce calling for conferences at all levels on the exchange of
materials (*wu-tzu chiao-liu hui*).[50] The establishment of this hierarchic system of conferences for concluding combined contracts institutionalized a
method of signing contracts that had been used frequently in the 1950s to
bring together and control exchanges between production, supply, and marketing units.[51]

In response to the desperate need to stimulate the production and exchange of sideline and small-scale agricultural and handicraft products, the
materials-exchange conference system is mainly directed toward third-category

goods. First-category goods cannot be dealt with by these conferences, and second-category goods can enter into the exchanges only insofar as they are not otherwise obligated for assigned sale to the state.[52]

The job of the basic-level materials-exchange conference—normally held on the *hsien* level—is to conclude concrete contracts implementing general contracts and agreements previously signed at conferences held on the special district, province, and central levels.[53] Thus, the county government organs, people's commune representatives, production-brigade representatives, and others who participate in the basic-level conferences ideally conclude contracts on the basis of the needs of larger divisions.[54]

While materials-exchange conferences were being called, the third step to stimulate the economy was the re-establishment of private plots.[55] The dispersed, small-scale production of private plots is amenable neither to direct state control through plans nor to efficient indirect control through contracts. Therefore, in September 1959, as the natural companion to the private plot, the setting up of the rural market was announced.[56] This final step completed the framework for a return to a pattern of commerce very similar, if more clearly and explicitly organized, to that prevailing in 1956 and 1957.

The rural markets are clearly intended to be complementary to state planning and the contract system.[57] Unlike the materials-exchange conferences, where contracts between units are used for exchanging products on a wholesale basis, the rural market is primarily for on-the-spot retail sales by individual peasants and production units.[58] The merchandise that can be handled in rural market trade in general is limited, in the case of first- and second-category items, to what is left over after completing state tasks (either in the form of plans and/or contracts),[59] and, in the case of third-category items, to what is left over after completing state tasks (set only for important items) and contracts.

As with the materials-exchange conferences, the main purpose of the rural market is to encourage the production and exchange of third-category products. Ideally, restricted in terms of merchandise, prices, participants and geographical compass, the rural market is an instrument for facilitating and minimally supervising the small-scale localized trade so necessary to the economy. Most rural market trade appears to be in these third-category commodities. Theoretically at least, contracts concluded at materials-exchange conferences inversely define the scope of rural market trade.

Delay in Implementation

Although this framework was set up in 1959, the policies regarding the rural markets, private plots, and materials-exchange conferences were not implemented effectively on a wide scale until early 1961. Whatever its causes,[60] the delay should not obscure the fact that within seven months after the high tide of the commune movement the regime was methodically preparing a structure for carrying out a policy of calculated retreat.

The retrenchment from the 1958 high tide of social mobilization continued at a slow pace in 1959 and 1960. A number of rural markets and materials-exchange conferences were held during 1959, but not on a nation-wide basis.[61] In early 1961, the great reorganization of the economy hit full stride. Coinciding with the reorganization of the commune[62] and the broad reestablishment of private plots, the SMCs were severed from the commune administration into which they had been merged in 1958.

During the first nine months of 1961, over 40,000 rural markets were held. Materials-exchange conferences were convened in at least a majority of the provinces at the provincial or special district levels. Correspondingly, the conclusion of contracts seems to have been widespread.[63] Under these conditions rural commerce began to improve, particularly in third-category products.

In the wake of the good harvest and upturn in 1962 and 1963, the number of large-scale materials-exchange conferences held in 1963 appears to have been greater than in 1962.[64] The number and value of contracts concluded at these conferences appear to have exceeded the figures for 1961.[65]

Another Perspective

The above analysis of the integrated relationship between the state, contracts, and the rural markets fits very well with G. William Skinner's more general analysis in his excellent pioneering study, "Marketing and Social Structure in Rural China."[66] Skinner postulates a hierarchical structure of marketing and related social systems in *traditional* China. Then, in examining the rural marketing structure in Communist China, he concludes that as of 1964 "there has been a remarkable continuity in the functions performed in each type of market town."[67]

Skinner's "standard market," which for present purposes may be taken as nearly equivalent to what I refer to as the rural market,[68] is the lowest level in the hierarchy. It was, and still is, a periodic market which meets "all the normal trade needs of the peasant household." The standard market "was the starting point for the upward flow of agricultural . . . and craft items into higher reaches of the marketing system and also the termination of the downward flow of items [originating in higher markets and] destined for peasant consumption."[69] Today, of course, many of the functions attributed to traditional standard market trade are performed supplementally or primarily by state agencies, warehouses, special trading companies, SMCs, and materials-exchange conferences. This displacement of the rural market proper is even greater today for markets above the standard market, the "intermediate" and "central" markets.[70]

The materials-exchange conferences (*wu-tzu chiao-liu hui*)—"commodities-exchange fairs" and "commodities-exchange meetings," as Skinner differentially translates it—"carry out functions which have been traditionally associated with the intermediate market town . . . to 'expedite the flow of commodities from wholesale to retail outlets.'" Referring to materials-exchange conferences *above* the basic level, Skinner writes that "they are devoted to transactions which contract for the future delivery of third-category goods in wholesale lots from one unit in the relevant city trading (or central marketing) system to another."[71]

I take exception to Professor Skinner's analysis only insofar as I am correct in inferring from his differential translation of the word *hui*—as "fair" for lower-level conferences and as "meeting" for higher-level ones—that he sees contracts as significant only at higher-level materials-exchange conferences. True, there is evidence to support the view that basic-level materials-exchange conferences may at times have deviated from the model of carrying out wholesale exchanges through contracts. In 1961, for example, numerous small-scale materials-exchange conferences, sounding very similar to rural markets and dealing primarily in on-the-spot retail trading, were held in Heilungkiang and Kwangtung.[72]

Nevertheless, it appears to have been the intention of the leadership when it established the materials-exchange conference system in 1959 to differentiate it from the rural market by designing the former as a hierarchical institution for level-by-level signing of contracts. If contracts are not signed at basic-level conferences, those signed at higher levels cannot be performed effectively. Moreover, unless contracts are signed at basic-level materials-exchange conferences, it is more difficult to restrict the scope of rural market trade. Therefore, if it is true that contracts do not predominate yet at basic-level materials-exchange conferences, I suggest that it is primarily a matter of time before they do, assuming policies remain unchanged.

Contradictions

Contracts in China involve intense contradictions between administration and production and between the government's desire to stimulate initiative at the lowest levels and its reluctance to relinquish significant authority. Since administration and authority dominate in China, these contradictions pose the fundamental issue of the role and importance of contracts. In the Soviet Union this issue seems to have been decided in the mid-1960s in favor of enhancing the significant use of contracts. Whether China in the 1960s has attempted to deal with its contradiction in a similar manner is a matter of dispute. I tend to think not. Professor Franz Schurmann, I take it, thinks that China has.

The Soviet Case

The Soviet pattern is to expand the real role of contracts in a broad range of less planned or unplanned dealings, while making it clear that the role of contracts in closely planned areas is near marginal. The pattern, in other words, is to encourage the use of contracts in areas where they may perform important and unique functions. Two fine papers on Soviet contracts suggest a two-fold trend within this overall pattern.[73] (1) In line with the 1964 reforms which liberalized the economy, broad areas of decision-making have been decentralized down to the level of the manager of the production unit. As a result, the importance of the market and of the private autonomy element in contracts, especially in the consumer goods sector, has been enhanced. (2) On the other hand, within the shrinking area of rigidly planned production and distribution, the extreme subordination of contracts to plans often relegates contracts to the realm of a formality or declaratory act. This is true, notwithstanding the fact that contracts may continue to perform certain functions even in this area.[74]

In situations where planning acts (that is, administrative acts) adequately define the relationship between potential contracting parties, the actual signing of a contract is taken to be unnecessary and "the parties are considered to have entered into contractual relations if neither notifies the other of his refusal to do so within ten days after receipt of the *nariad*."[75] This procedure, called "taking the *nariad* to completion," does not represent a substantive decrease in the sphere of private ordering. Rather, it reflects a recognition that procedures appropriate to private ordering formerly had been employed in situations where they had no substantive significance. As a result, today it is recognized that civil law obligations between parties can arise on the basis of administrative acts alone.

Contract terms which duplicate or violate prior or subsequent plans or planning acts are superfluous. Thus, State Arbitrazh has held that a contract need not incorporate all the terms normally required for completion so long as those left out are embodied in planning acts. And, again if certain terms of a contract are nullified for violating plans, the contract as a whole is not thereby invalidated so long as the terms, if essential, are incorporated in plans. Similarly, the terms of new planning acts that are contrary to the terms of previously-concluded contracts govern the relation between the parties to the contract even if the parties did not amend their contract to incorporate the new terms.[76]

On the other hand, within the economy as a whole, it appears that "the relative weight of contract *vis-à-vis* plan is growing."

> Managers today enjoy far more flexibility than they did ten years ago . . . plans are less detailed than formerly, and jurists are calling for even less detailization; fewer goods are being distributed in a centralized manner; buyers can refuse to contract for unnecessary goods allotted to them by *nariad;* retailers can unilaterally rescind contracts of delivery if consumer demand declines.[77]

Many manufacturers and retailers have been given greater freedom to determine with whom to buy and to sell, and at what price. Moreover, there appears to be a consensus among Soviet legal scholars that "in the future, with greater availability of goods, the number of contracts directly planned will decline."[78]

The Chinese Case

The reorganization of the Chinese economy in the early 1960s, as Schurmann points out, reflects "an interplay" of three "conflicting tendencies":[79] (1) centralization; (2) transfer of decision-making powers downward to lower-level *administrative* units (what Schurmann calls "decentralization II"); and (3) transfer of decision-making powers downward to *production* units (what Schurmann calls "decentralization I").[80] I disagree with Schurmann on the significance of the third tendency, decentralization I. Leaving aside centralization, I tend to believe that decentralization II, rather than decentralization I, has dominated the reorganization. Schurmann characterizes the balance among the three conflicting tendencies more in favor of decentralization I, arguing that under the reorganization, the contract system generally incorporates significant independent operational autonomy for the production unit.[81] Schurmann, then, in my terms, is arguing that the contradictions in contracts have been resolved, to an important degree, in favor of the production unit and at the expense of administrative authority.

From my research in the sphere of agriculture, I view the resurgence of contracts in the 1960s within the context of administrative decentralization, because it has been accompanied by a strengthening of institutional controls, namely, the materials-exchange conferences. Professor Schurmann appears to be aware of this role of the materials-exchange conferences.[82] However, he may interpret such conferences and analogous institutions in industry primarily as facilities for implementing decentralization I.

Unlike Professor Schurmann, I am hesitant to ascribe significant individual initiative to parties to industrial contracts. On the basis of my admittedly limited knowledge of industrial transactions, let me suggest that Professor Schurmann may have overstated somewhat his case for what he calls direct contracts between production units—that is, contracts made more or less outside of state channels.[83] My reservations are two-fold, relating to the legal framework required for direct contracts and to the level of information about "market conditions" necessary for such contracts.

First, if the area for negotiation between parties to an industrial contract had been significantly enlarged since 1961, the need for sophisticated contractual rules to provide a framework for individual initiative would have been heightened. Yet unlike the Soviet Union whose response to its NEP period was to promulgate codes, China has not to our knowledge done so.[84] Moreover, after a brief flirtation with a state arbitration system similar to the U.S.S.R. State Arbitrazh, China abandoned the experiment as unnecessary. It appears, then, that China does not have an institution capable of resolving contract disputes in such a manner as to supply the uniformity of interpretation and refinement of reasoning necessary to provide a suitable framework for individual initiative. In fact, it is not clear to me what existing institutions in China would enforce direct contracts. This is a vital question, because agreements that are not institutionally enforced cannot function as contracts.

Second, to work at all efficiently, direct contracts would require facilities to provide a reasonable level of information about conditions of supply and demand. But it appears that the use of advertising, and concomitantly the scope of direct, spontaneous exchange relationships, may be limited to surplus goods, as an interview of mine suggested. Schurmann is aware of the use of advertising to dispose of surpluses, but he clearly does not see it as limited to such goods.[85] Without extensive advertising, the widespread use of Chinese *tolkach*,[86] or other functional equivalents,[87] the limitations on direct exchange relationships outside of state channels are apparent.

If I find Schurmann's emphasis on the role of production-unit initiative and direct contracts somewhat overdone, I would accept his description as a *prediction* of what the future probably holds for China as its economy becomes increasingly complex and efficiency becomes more important. With this view toward the future it is useful to recall that "contradictory elements . . . , though they create conflict, . . . also provide the leaders with multiple possibilities for action."[88]

Conclusion

The Chinese, to my mind, have not resolved the contradictions inherent in contracts within their system. Unlike the Russians, who have expanded the significant use of contracts by allowing greater latitude for individual initiative at the production-unit level, the Chinese generally have tended to limit the decentralization of decision making, insofar as it is relevant to contracts, to the administrative structure. In agriculture in the 1960s they strengthened the administrative structure encompassing contracts by setting up the materials-exchange conference hierarchy.

While it is always imprudent to emphasize the high degree of actual control in such a complex, contradictory, and inefficient system, I am inclined to minimize the importance of individual initiative at the production-unit level in most contracts in China. In practice such initiative appears to have been severely restricted by various organizational and ideological factors. If at a particular time it was not the central government, then it was the provincial or lower-level government, or, if not the government, then the Party, which performed the restraining function. There seems always to have been someone.

This is not, however, to accept the simple commandist view of contracts or of Chinese society. In the first place, commands, if they are to function with minimal efficiency over time and on a society-wide basis, must incorporate elements of agreement. For authority to be accepted over time, it must reflect to some degree the wishes and capacities of those over whom it is being exercised.[89] Thus, although the Chinese government as a party to a contract is in a superior bargaining position, there still remains an element, if only implicitly, of agreement. Moreover, in the nature of things inevitably there are lacunae between "commands" in which individual initiative will be important, though it need not always be embodied in contracts.

Chapter V

CONTRACTS IN AMERICA IN PERSPECTIVE

This book began with the inchoate recognition that an analysis of contracts in China by a Western-trained lawyer is likely to be biased by comparative perspectives derived almost unconsciously from the Anglo-American tradition and mythology of contracts. The ideal type of contract in the West,
maximizing individual freedom and minimizing governmental interference, a
contract closely associated with the free market, seems largely out of place
in a socialist, planned economy.

Yet, as Chapter II indicates, the Chinese have praised and used contracts
widely, and their contract law "on the books" has been adapted to their economy (Chapter III). Nevertheless, the initial suspicion that contracts somehow
are out of place in Communist China, that they are not very significant, seems
to be confirmed in Chapter IV, where the ways in which contracts are tightly
integrated into the rural economy, the severe restrictions on the autonomy
of parties to contracts, and the very limited significance in practice of technical rules of contract law all are made manifest. If the Western classical ideal
is the abstract contract, in Communist China it appears the ideal is nearly the
reverse: in China the nature of the goods or services contracted for and the
nature of the contracting parties are foremost concerns of Chinese public policy and are to a substantial degree determinative of the particular way a type
of contract will be made and handled.

But is not the major underlying point here simply that the Chinese reality is qualitatively different from the classical Anglo-American ideal and from
our practice as reflected in case books? Today in American practice also, as I
argue below, societal concerns other than those reflected in classical liberal
thought and practice tend to dominate contracts. The ideal of the individual
bargain as the heart of contract has been largely eroded in America by developments as diverse as the use of collective bargaining and standardized contracts.
The scope of the contract residuary, discussed in Chapter I, has been shrunk. In
fact, even the significance of the legalistic aspects of contracts today seems
quite limited for a very substantial portion of American businessmen. Moreover,
the modal pattern in America of types of contracts and how they are formally
"processed" appears to be shifting toward a Chinese-type administrative practice and away from the liberal ideal. But, then, what is left of contracts? Why
and when do Americans and Chinese employ contracts? And what do these patterns imply about the role of contracts in development? Only tentative answers
can be provided to these questions, since evidence is scanty and speculation just
that.

A Developmental Theory of Contracts

Perhaps the place to begin this reevaluation of contracts in the West and in China is by spelling out a general theory of socioeconomic development in which the role of contracts as seen by liberal ideology is placed in a longer-term perspective. And, perhaps, the best way to begin that task is with a brief critique of Maine's theory of historical progress which sees societal development in terms of the movement from societies that predominantly relied for ordering relations between men on status to those that predominantly rely on contract.

Maine's theory, whether or not he meant it so broadly, has been taken to be a statement about the development of societies as a whole. Seen as such, there are at least two major flaws in the theory that have misled many subsequent analysts.

First, the significance of the theory must be restricted in two senses: (a) its validity seems strongest with regard to how relationships are initiated, and weakest regarding the implications of relationships once entered into; and, excepting that more limited validity, (b) Maine's theory does not appear to describe a secular trend in history, but rather a phase of history which seems more intelligible in cyclical terms than in terms of a straight-line unilinear progression.

Second and relatedly, Maine's theory is itself vulnerable to analysis in terms of sociology of knowledge. His theory is blatantly historically and culturally relative, reflecting the individualistic biases of liberal ideology. Because it emphasized what had changed, it failed to take account of what had remained substantially the same. More precisely, Maine's theory focusing on contract contributed to the kind of dichotomous analysis that takes contract and status as separate, rather than as intimately related, phenomena.[1]

If I were to propose an alternative vision of development, its broad outlines would suggest a cyclical or spiral "development" from (1) a status-dominated society, which nonetheless cannot be understood wholly in terms of status considerations, to (2) a contract-dominated society, in which rapid flux tends to undermine the remnant and changing content and significance of status, to (3) a society in which there continues to be an increase in the absolute number of contracts made and in which contract and status obviously are interacting, but in which the content and significance of both are moving in the direction of increased status implications.

Roscoe Pound was one of the first to recognize the relativism of Maine's argument and to reassert the importance of legal status in Anglo-American society.[2] Maine saw the individual and society as being in conflict. He saw the end of law, on its negative side, as being to remove societal obstacles to individual self-assertion and, on its positive side, to spread and enforce contractual obligations, which were taken as the embodiment of maximum individual self-assertion. Maine, like other nineteenth-century writers, stressed the individual will and tended to ignore a core element of Anglo-American common law, the legal concept of relation. The concept of relation, a phenomenon not wholly manipulatable by the individual will, appears to be the irreducible minimum recognition a legal system must accord "status." Pound, putting the matter somewhat differently, writes:

> In the Nineteenth Century the feudal contribution to the common law was in disfavor. Jurists thought of individuals and contracts rather than of groups and relations. The conception of the abstract individual ruled in legal philosophy. The medieval guilds were gone and the legal position of trade unions and a legal theory of collective bargaining had not yet become problems for the lawyer. Hence, the Nineteenth-Century lawyer thought ill of anything that had the look of the archaic institution of status.
>
> [But] the idea of relation, and of legal consequences flowing therefrom pervades every part of Anglo-American law. . . The common-law lawyer . . . [for example], thinks of the relation of principal and agent and of powers, rights, duties and liabilities, not as willed by the parties, but as incident to and involved in the relation. . . When the relation does exist . . . a train of legal consequences follow. . .[3]

Maine's generalization, Pound writes, in polemical exaggeration,

> has no basis in Anglo-American legal history, and the whole course of English and American law today is belying it, unless, indeed, we are progressing backward. If it be said that statutes restricting freedom of contract between employer and employee are a legislative phenomenon . . . out of line . . . [with] the common law, one may point to the law of public-service companies or to the law of insurance or to the law of surety companies. In each case, and these are relatively recent judicial developments, . . . the common-law idea of relation and of the rights, duties and liabilities involved therein, has prevailed at the expense of the idea of contract.
>
> Even more significant is the legislative development whereby duties and liabilities are imposed on the employer in the relation of employer and employee, not because he has so willed, not because he is at fault, but because the nature of the relation is deemed to call for it. Such is the settled tendency of the present, and it is but a return to the common-law conception of the relation of master and servant with reciprocal rights and duties and with liabilities imposed in view of the exigencies of the relation. . . It is not out of line with the common law to deal with causes where the relation of master and servant exists differently from causes where there is no such relation. On the contrary, the Nineteenth Century was out of line with the common law when it sought to treat

the relation of master and servant in any other way. . . In the pioneer agricultural societies of Nineteenth-Century America such a conception sufficed. In the industrial and urban society of today classes and groups and relations must be taken account of no less than individuals.[4]

The concept of relation, Pound writes, is "the characteristic common-law mode of treating legal problems which we derived from the analogy of the incidents of feudal tenure." Presciently, Pound concludes that the concept of relation is a critical legal tool for "making our received legal tradition a living force for justice in the society of today and tomorrow."[5]

Thus, in the very categories of legal analysis that of necessity generally involve relational implications, types of status are preserved. Maine, focusing on how in industrializing societies relations are formed and, to a lesser extent, on how they are terminated (more or less "voluntarily" in the case of contracts as compared to more traditional modes), ignored this substantial continuing role of status incidents. For, once a relationship is entered into, the law frequently will impose incidents taken to be appropriate to that sort of relationship.

To the extent that abstract contract theory and nineteenth-century practice imposed *no* relational incidents other than those embodied in the contract itself, the nineteenth century might be taken in retrospect to have been something of an historical aberration. Such a restrictive approach to the institutional role of relations would imply that the relevant societal concept of justice in the nineteenth century had been reduced to bargain justice, pure and simple.[6]

But, as suggested above, if the greatly enhanced role of abstract contract in the nineteenth century is seen in terms of a relative rather than an absolute shift from status to contracts, then this phenomenon may be best understood as a phase of a cycle. To understand the cycle, these developments must be seen in a broader framework of social change than is provided by an exclusive focus on the legal institution of contracts. In this broader framework the intimate relationship over time and at any one time between contract and status becomes clearer.

The framework for analysis is one expressing degrees of standardized relations. Within such an inclusive framework status and contract become not dichotomous states but poles on a continuum. The criterion is the degree of standardization, with "pure" contracts representing highly individualized relations and "pure" status representing thoroughly standardized relations. In between the poles lie many twentieth-century contracts in which, like ancient family relations (status), peculiarities of the individual agreement are ignored, played down, or overridden and the contract produces a more or less standardized relation. The similarity between an ancient status relation and, say, a relation arising from the purchase today of a standard insurance policy is that

> a relation results in which the details of legal rights and duties are deter-mined not by reference to the particular intentions of the parties, but by reference to some standard set of rules made for them. In origin . . . [today's insurance relations] are, of course, contractual, [but] in their workings, they recall the regime of status. Maine's original statement has reference to a classification on the basis of origins. His argument ap-plies—and is generally applied—to a classification of relations on the ba-sis of their workings. In this sense, the difference between status and contract is not one of kind, but one of degree; and in this sense there has clearly been a long-enduring tendency in English law from status to con-tract, and—in the last two generations—an equally distinct veering back to status.[7]

This shift in the last several centuries from highly standardized relations to highly individualized ones and then back to more standardized relations (status to contract to status) suggests "a kind of pendulum movement back and forth between periods of standardization and periods of individualiza-tion."[8] Similar movements occurred previously in history, for example, in the development of feudalism. At first, feudalism itself reflected a shift from pre-feudal status to feudal relations, at inception based on contract. Later feudal relations, in turn, hardened into status.

It seems clear that highly standardized legal relations are a product, in part, of societies that are more or less active in imposing particular public pol-icies or concepts of justice on its members (the United States today or Com-munist China, for example). Highly individualized legal relations, on the oth-er hand, are a product of governments imbued with a philosophy of *laissez-faire* and committed primarily to the notion that justice is what the parties to the relation say it is. One would expect, then, that the more active a govern-ment is in facilitating real (as distinct from legal) equality, and the more ac-tive a government is in insuring that certain minimum standards are realized in economic dealings, the more likely is it to emphasize standardized rela-tions. This notion of social enfranchisement through status law is implied in the more general and well-known distinction in political theory between nega-tive and positive freedom[9] and in the more specific legal analysis of Nathan Isaacs, who writes:

> The movement toward status law clashes, of course, with the ideal of individual freedom in the negative sense of "absence of restraint" or *laissez-faire.* Yet, freedom in the positive sense of presence of opportu-nity is being served by the social interference with contract. There is still much to be gained by the further standardizing of the relations in which society has an interest, in order to remove them from the control

of the accident of power in individual bargaining. . . [T]he current of legal development . . . [is] resorting to status as an instrument at this particular time for the further enfranchisement of those to whom freedom of contract has become a mockery. Freedom of contract is not synonymous with liberty, nor is status slavery . . . the knife can cut both ways. In . . . the 1300's, [for example,] status law was being used to drive laborers to their work; now it is looked to to force employers to a realization of their social duties. It then practically created a maximum wage; today it is the messenger of a minimum wage.[10]

Implicit in this development theory is the idea that, as societies have stabilized and systematized their legal structures and theories, standardized legal relations have tended to become dominant.[11] Whereas in times of very rapid change and innovation in the units and forms of doing business, more individualized legal relations have tended to become more significant. It hardly seems mere happenstance, for example, that the zenith of the impact of contracts in England and America generally coincided with the industrial revolution in each country, periods of unprecedented rapid social change.

More on Status and Contract

Such a developmental theory of the cyclical relation between legal status and contract, however, suggests only one aspect of their complex relationship. It tends to slight the continuing process of *interaction* between status and contract over time and at a single time. The history of the Anglo-American development of contract law is a history of their interaction over time.

In the fifteenth century as the concept of purely promissory liability developed in the shape of new forms of action, such as trespass on the case and special assumpsit, status considerations provided the basis for moving to contractual liability on the promise. The action of trespass proper did not lie for damages caused by a defendant in cases where a plaintiff-owner had voluntarily entrusted property to the defendant. The development of trespass on the case to cover such situations emerged first in cases where "the defendant, because of his trade or calling, undertook to handle or deliver plaintiff's body or goods safely and skillfully." In those cases, "failure on . . . [the defendant's] part to live up to the standards of his profession, *i.e.,* status, gradually became negligence sufficient to maintain an action in trespass on the case. Thus, [arose] the first cases in which the breach of what we now call a contractual agreement was penalized."[12]

Subsequently, the scope of trespass on the case was expanded to include any undertaking (for example, a bailment) that resulted through malfeasance in damage to the plaintiff, whether or not the defendant was exercising a common calling.

With this development the common law moved one step closer to recognition of contractual liability, *i.e.*, liability based on agreement rather than on status.

But the solution could not stop there. A person needed protection not only . . . [from] malfeasance . . . , but also . . . [from] nonfeasance. . . Again . . . [r]ecognition of liability . . . [for] nonfeasance . . . had its origin in status: it first came to be recognized as against persons who exercised a common calling. They were bound to serve those in need of their services.[13]

Thus, a blacksmith who unreasonably refused to shoe a horse upon request was held liable for damages. "To be sure, the smith had not broken a contract; he had merely refused to enter into what we now call a compulsory contract, into which a person of his status was obliged to enter."[14] Again, the status-like justification for liability developed into a contractual liability for nonfeasance.

Another example of interaction between status and contract is provided by the early defective automobile cases in the U.S. In the twentieth century, tort law—a status-like law which imposes rights, duties, and liabilities on individuals because of their relation to others—was instrumental in stimulating the broadening of contractual liability. When, because of the doctrine of privity of contract, the ultimate consumer of a defective product was unable on the basis of the purchase contract to directly sue the manufacturer, who was the dominant figure in the marketing chain, tort law stepped in and provided a direct remedy by imposing on the manufacturer a duty running to the ultimate user. Subsequently, several jurisdictions, accepting the direct action idea, allowed direct suit on the contract under implied warranties.[15]

But in a still broader sense contract and status are in constant interaction in the process of judicial enforcement of contracts. For, to understand the meaning of a particular contract, one must understand the socioeconomic context in which it is embedded, a context composed of myriad status-like relations. The importance of this context can be seen by briefly examining the traditional distinction between express, implied-in-fact, and implied-in-law contracts. Traditionally, the first two types of contracts are lumped together as based on the intention of the parties. Within a tradition that emphasizes the volition of the parties and holds that courts do not "make contracts for the parties" but merely give effect to their intentions, the distinction between express and implied-in-fact contracts is held to be insignificant. The only difference, not relevant to the pivotal role of intention itself, lies in the manner in which intention is manifest—in the case of express contracts intention being, by definition, express, and in the case of implied-in-fact contracts intention being derived from the circumstances of the case. Implied-in-law contracts (quasi-contracts), on the other hand, are distinguishable as imposed by law, the product of a legal fiction based not on the parties' intention but resorted to by courts in order to do justice in cases of unjust enrichment.

This sort of bifurcated traditional analysis blurs the institutional aspects of *all* contracts. As Kessler and Sharp point out:

> The techniques developed in dealing with implied-in-fact contracts help us to a better understanding of the social structure of contract in general. Confronted with the issue of whether or not to imply a contract, courts are faced with the complex task of finding objective criteria for determining the intention of the parties in the light of the circumstances surrounding the transaction. A court that wants to avoid the arbitrariness of "random behavior," so as not to disappoint the reasonable expectations of litigants and in order to make future decisions predictable, has to look, whenever possible, to the facts of general business experience and understanding. And their experience with implied contracts has helped courts to realize that the task of interpreting and enforcing an express contract is not fundamentally different. The meaning to be given to all modes of expression of a contractual intent is found by a process of implication and inference. In this sense, all contracts are implied contracts." . . . [And, as Durkheim wrote,] "the normal contract is not an isolated act, but an incident in the conduct of business or in the framework of some more general relation. . . It will frequently be set against the background of usage familiar to all who engage in similar negotiations and which may be supposed to govern the language of a particular agreement."[16]

Moreover, there are other problems with the traditional distinction, based on volition, between contracts-in-fact and those implied-in-law, problems which in a different way reflect a slighting of the contextual or extracontractual relational elements in many transactions. In this regard Corbin has written that "considerations of equity and morality play a large part in the process of finding a promise by inference of fact as well as in constructing a quasi-contract without such inference at all. The exact terms of the promise that is 'implied' [in fact] must frequently be determined by what equity and morality appear to require after the parties have come into contact."[17] Thus, on closer examination, it becomes clear that

> the "implied contract" and the "implied term" are tools enabling a court to impose and adjust contractual liability in accordance with the demands of the cultural setting. . . The degree of control over private volition thus exercised is in some measure obscured by the fact that the courts habitually speak of "interpretation" even if their task of enforcing a contract has led them far beyond a mere determination of the meaning of the symbols of expression used by the contracting parties.[18]

Status (legal and non-legal) and contract, thus, constantly are in interaction defining and modifying each other.

In two further senses the importance of contract in America as a highly particularized, purely voluntary, legal relation is even more restricted than is suggested by this interaction. In the first place the scope and depth of standardization today is great. While I have adverted to the degree to which contractual relations have become standardized, my focus has been almost exclusively on standardization imposed by government as an instrument of public policy, and even in that realm the extent of standardization has only been hinted at. In the second place, legal relations (for example, contractual relations) are only one component in the total relationship between the parties, a component that frequently, depending upon the nature of the transaction and the quality of the parties' non-legal relationship, may be quite unimportant.

What's Left of Freedom of Contract?

At least four major factors in American socioeconomic development over the last seventy-odd years help to explain the increasing standardization of contracts in law and practice.

The first is the widespread process of concentration in industry and business, corresponding to an increasing urbanization and standardization of life. Its legal result is "standard contract," or "contract of adhesion." The second factor is the increasing substitution of collective for individual bargaining in industrial society. Its legal product is the collective contract between management and labor, with a varying degree of state interference. The third factor is the tremendous expansion of the welfare and social service functions of the state. . . Its legal effect is twofold: on the one hand, it has led to a multitude of statutory terms of contract, substituted for or added to the terms agreed upon between the parties; on the other hand, it has led to a vast increase of contracts in which government departments or other public authorities are on one side, and a private party on the other. The effect of this on the law of contract, though . . . little explored, is profound.

All these developments affect the . . . practice of contract. . . The ideology of social security means emphasis on stability and a corresponding lack of mobility [that is, the free making and unmaking of legal relations through contract], especially in employment contracts. The standardization of contract greatly restricts the freedom of the weaker party, and is usually accompanied by inequality of bargaining power. Collective bargaining, on the other hand, has substantially restored equality of bargaining power between employers and employees, though increasingly at the cost of individual freedom, as the legal or practical compulsion

to join employers' associations and trade unions progresses. The imposition of statutory duties in the interest of social justice largely sacrifices
mobility for stability and security. The increasing participation of public authority in contract creates the wider . . . problem of the dual
function of the state, as a superior and as an equal. . .

[Regarding] the aspect of . . . [the scope of] economic risk which it
is the function of the law of contract to safeguard . . . , the elaboration
of remedies for breach of contract . . . is increasingly affected by the
spread of such political, economic, and social upheavals as war, revolution, or inflation. The legal result is the doctrine of frustration of contract, with its consequent extension of legal excuses for the non-performance of contract.[19]

With regard to the fantastic degree of standardization that has occurred
in twentieth-century America, at least five somewhat overlapping and related
forms of standardization are worth mentioning: (1) compulsory contracts;
(2) contracts in which certain terms are compulsory; (3) collective bargaining
agreements; (4) standard form contracts; and (5) contracts with government
agencies (especially contracts with the federal government).

(1) If, as Max Radin has written, we tend to view status primarily as the
complex of legal obligations imposed by law on men who neither voluntarily
assumed them nor can easily shake them off nor modify them by a simple act
of will, then "compulsory contracts" may seem to be more in the nature of
status than of contracts.[20] This is particularly true of the pure form of compulsory contract in which the contract is required to be made and its terms
largely determined by legal compulsion (for example, public utility contracts). It is less true in the diluted compulsory contract, in which there is no
duty to contract with all comers but in which once one decides to make a certain type of contract (for example, an insurance contract) most of the terms
of the contract are imposed by law rather than created by bargain and agreement.

The pure form covers much of the huge and expanding field of public
services, such as the utilities and the transportation and communication net-
works. Businesses in these fields generally may not freely choose the parties
with whom they will do business nor freely determine the terms on which
they contract to do business. Generally speaking, they may not, on penalty
of forfeiting the right to engage in such business at all, voluntarily stop making contracts with members of the public. And in those contracts they do
make, terms regarding rates and other central rights and duties generally are
determined by public authorities. Although this sort of compulsory contract
is not an innovation of industrial society—it existed also in the Middle Ages—
one critical development of the twentieth century has been the extension of
the concept of a duty of public service to new categories of businesses and

the setting up of administrative agencies to supervise the operation of businesses so imbued with the public interest.

Similarly, the banking and insurance businesses are closely supervised and controlled by public authorities, in part through the diluted compulsory contract. While bankers and insurance companies generally are under no obligation to contract with every member of the public who wishes to do business with them, once they do enter into such contracts, their capability to shape the agreements individually is very limited. The special sanctions provided for this sort of contract are reminiscent of similar sanctions in socialist countries, applicable there more broadly to all business contracts in planned sectors: "The making of a non-conforming [insurance] contract [in the U.S., for example,] is a penal offense; and the non-conforming contract is to be read as if it contained the prescribed terms, and no others inconsistent with them."[21]

(2) Less restrictive of the individual bargain than the compulsory contract but much more widespread and of greater direct impact on contracts and the economy as a whole is the category of compulsory terms. These terms, frequently the by-product of statutory duties arising from a relationship between the contracting parties, such as employer-employee or landlord-tenant, prescribe a part of the bargain. The by-now-familiar battery of such terms includes terms affected by minimum wage legislation, rent control acts, and workmen's compensation laws.

(3) Allied to the increasing use by government of the technique of compelling certain terms to be in contracts has been the increasing use of collective rather than individual contracts. Both developments reflect attempts to change the relative distribution of the power to determine the nature of the contractual bargain, the latter development, at least at the outset, being more a "private" effort. The growth of corporations (management's version of the collective) and labor unions has been part of a long-term social process in which group bargaining came to displace individual bargaining. But the sudden growth of collective bargaining in the U.S. in the 1930s and 1940s was largely the product of the 1935 National Labor Relations Act, which required employers to bargain collectively with representatives selected by their employees. Collective bargaining has meant that

> firstly, . . . [the contract arrived at through collective bargaining] resembles a standard contract of business and industry in that standardized terms regulate the conditions of employment of millions of individuals. Secondly, it is a most important instance of a legislative function delegated . . . from government to social groups. . . . That the modern collective agreement has a social and economic significance different from that of the common law contract is certain. It establishes, in the form of group agreement, a code of conduct between powerful sections of

the population controlling the vital factors of national production . . . [Such contracts] . . . not only involve the groups concerned but also the public. Hence the need for a varying degree of state interference. . . Thirdly, the freedom of the individual to bargain on his terms of employment is inevitably curtailed by the prevalence of collective bargaining. It is even excluded where the "closed shop" is recognized, either legally or *de facto.* Fourthly, this lack of freedom is compensated by a substantial restoration of equality of bargaining power. It is not [, however,] the individual employee who has . . . gained equality, but the trade union negotiating on his behalf.[22]

(4) In addition to the pressures toward standardization of contract exerted by governments and the public-private pressures reflected in the development of institutionalized collective bargaining, various segments of the business world itself frequently have standardized contracts covering transactions routinely found within a particular segment. Strange in view of the semantic contradiction, but illuminating in terms of the interaction between seeming opposites in history, standardization in the context of mass production has operated as a mode of specialization. The movement in various fields of business toward standard or form contracts grew out of a recognition that the general law of contract, which attempted to deal homogeneously with different *sorts* of contracts, was too general, that, as such, it too often frustrated commercial expectations, and that it therefore required tailoring to fit commercial practice. Within the area of suppletive or yielding contract rules, the parties could create a standard contract to fit a recurring type of transaction.

If the standardized contract reflects the impersonality of the market, it equally reflects the crystallizing of market exchanges into patterned exchanges. While this form of contract obviously is not so "specialized" as the highly individualized "freely-bargained" contract, standard contracts involve a degree of type-standardizing commensurate with the concomitant standardization of specialized production processes and goods. Thereby, bargains (or many aspects of bargains), like goods, are mass produced. The alternative, "the working out of thousands of individual contract terms for substantially similar transactions would be as uneconomical as the use of antiquated machinery."[23]

The results of such mass production are two-fold. First, the processes of negotiating, concluding, executing, and supervising contracts are simplified and probably made less costly, potentially benefiting the ultimate consumer. But, second, as individual bargaining is severely limited or even eliminated by means of such standardization, and as the parties to standard contracts rarely are in equal bargaining positions, the standardized contract has frequently been a device for imposing one party's will on the other. In these cases the terms are more or less dictated on a take-it-or-leave-it basis. Thus, standardized contracts frequently are contracts of adhesion, contracts which present problems of fairness and reasonableness—what Llewellyn refers to as "the minimum decencies . . . [upon] which a court will insist . . . as essential to an enforceable bargain of a given type"—problems not easily resolved or even conceptualized in terms of purely abstract contract theory.[24] And the impact of standard contracts, products of prospectively agreed-upon or unilaterally imposed general conditions which are to obtain in an industry or line of trade (unless government or opposing, powerful, private groups interfere), can hardly be exaggerated.

In modern society a member of a community, for example, may be said at best to be free to choose which among available contracts he wishes to make, but he is hardly free to choose whether or not to make contracts at all. The economic organization of modern society creates a compulsion in this respect, as it does, to a lesser degree, regarding the choice of contract terms:

The vast majority of . . . persons in any given community must make contracts in order to live. They have a considerable latitude of choice in the large centres of population. But in smaller centres and in rural areas, that latitude does not exist. In a steel town or a mining town, the vast majority must work in these commodities or not work at all, which is another way of saying that they have no choice in the matter.

It is quite true that in the United States the population is probably more mobile than anywhere else, so that there are more opportunities for the free exercise of choice of economic function than is normally the case in modern states. This choice involves of course a freedom of selecting the contracts one wishes to make. But even in the United States this freedom is limited by the difficulties of migration, social, economic and personal. A very large portion do not really have it, and in any case soon lose it, since the opportunities offered by migration are in inverse proportion to the age of the migrant.

Not only is there [, in this sense,] a compulsion in the making of the contract, but the terms of the contract are not as much the creation of the free consent of the contracting parties as we like to believe. *Almost all the contracts that we must make to secure and maintain our living are type contracts.* They are standardized as to performance and as to consideration for performance. The range of return from investments is limited and does not deviate very much at any given time. . . *It would be very difficult, as a matter of fact, to select any type of economic activity that involves a substantial number of persons, in which one could not tell in advance what the terms of the contract are likely to be which these persons are compelled [by their socio-economic context] to make.*

[T]he de facto monopoly of huge combinations . . . limit[s] the range of free contractual activity for all but a small number of persons . . . [As to two groups of commodities especially, those] in which a special market has been created by extensively advertising under a brand-name or label . . . [and those] in which technological advances have become the exclusive property of a few powerful bodies . . . , the net result . . . must be that those who wish to deal in the products so advertised and so controlled . . . must do so on terms which they have no voice in fixing and no bargaining power to modify.[25]

In this perspective the practice of contracts today in the United States provides a caricature of the ideology of "freedom of contract":

With the decline of the free enterprise system due to the innate trend of competitive capitalism towards monopoly, the meaning of contract has changed radically. Society, when granting freedom of contract, does not guarantee that all members of the community will be able to make use of it to the same extent. On the contrary, the law, by protecting the unequal distribution of property, does nothing to prevent freedom of contract from becoming a one-sided privilege. *Society, by proclaiming freedom of contract, guarantees that it will not interfere with the exercise of power by contract. Freedom of contract enables enterprisers to legislate by contract and, what is even more important, to legislate in a substantially authoritarian manner without using the appearance of authoritarian forms.* Standard contracts in particular could thus become effective instruments in the hands of powerful industrial and commercial overlords enabling them to impose a new feudal order of their own making upon a vast host of vassals. This spectacle is all the more fascinating since not more than a hundred years ago contract ideology had been successfully used to break down the last vestiges of a patriarchal and benevolent feudal order in the field of master and servant. . . Thus the return back from contract to status which we experience today was greatly facilitated by the fact that the belief in freedom of contract has remained one of the firmest axioms in the whole fabric of the social philosophy of our culture.[26]

(5) Another major cause of standardization in America since the 1930s has been the growth of government, especially federal government, contracting: *"Since 1940 the Federal Government has been the single best customer of private enterprise."*[27]

In one respect federal government contracts can be seen simply as a subcategory of standard form contracts. Because of both the bureaucratic quality of government administration, which requires certain standard terms, and the same pressures as exist in the private realm to standardize contracts according to recurrent types of transactions, many of the terms of government contracts tend to be "boiler plate provisions," which the private businessman contracting with government generally must accept willy-nilly if he wishes to contract at all. To the extent such government-written standard provisions pervade the agreement, government contracts are the product not of bargain but of adhesion. In this respect they are similar to highly standardized contracts in the private sphere in which a dominant party more or less imposes predominantly self-serving terms on the weaker party.

Nevertheless, despite parallels to a private party in a superior bargaining position,[28] the government appears to tend not to assert its superior position. Thus, a "basic idea of federal procurement is to give the contractor a decent profit."[29] The government, as an ally of big business, has no special desire to drive a hard bargain as to price. Similarly, a tone of fairness seems to permeate standard government forms, providing prospectively for such matters as compensation for government acts which, while technically not violations of contract provisions, are disadvantageous to the private party, and providing somewhat formalized, due-process-oriented administrative dispute-settling techniques, designed in part to protect the interests of private parties.[30]

But, on the other hand, in many respects the imposition by government on the weaker private party is much greater than in parallel private contracts, because the government as a party to a contract has broader concerns and capacities, and because it acts to protect those concerns. The government, for example, frequently imposes terms expressing general public policy not strictly relevant to the economic side of the transaction, terms protecting the integrity of government, and terms endowing government agents with qualified powers to unilaterally modify or terminate contracts without the agreement of the other party. Moreover, despite procedural safeguards to protect the interests of the private party, the range of judicial remedies practically available to the private party appears to be more circumscribed than in nongovernment contracts, and government control of contract performance and the sanctions government may wield greatly exceed comparable examples in the "private" economy. Government contracts in the United States function, in short, both as regulatory devices (that is, as an extended form of government administration disguised by the at least formally voluntary nature of the act by which the other party assumes its contractual duties) and as procurement devices (with material and other incentives built in). Government contracts tend to be determined by the government's assertion of its authority and one-sided power, with the significance of bargain reduced essentially to those terms about which the government is less assertive.

Let me be more specific. As a matter of public policy the government, for example, always or frequently demands in its contracts that: (a) supplies for public use generally be purchased from American manufacturers and be American-produced materials; (b) contractors not practice discrimination in their employment practices; and (c) certain government-set standards regarding wages, hours, and working conditions be adhered to by all contractors (administrative "black-listing" is one available sanction for contractors who violate such standards).

To protect government integrity, "gratuities" clauses, covenants against contingent fees, and the like are imposed on contractors with the government. A contractor who violates such clauses by trying, for example, subtly to bribe a government official to favor him in a contract award is liable to have the resulting contract terminated and is further liable for exemplary damages (penalties of "not less than three nor more than ten times the costs incurred by the contractor in providing the gratuity"), as well as for compensatory damages.[31]

Many types of government contracts, moreover, almost routinely include terms which endow the government with very significant, if qualified, unilateral, discretionary powers to alter or terminate the contract. The "changes clause," for example, empowers a government contracting officer to issue certain change orders which, within a specified range, modify contract performance, whether or not the contractor consents to the particular order and even if he opposes it. The contractor, in effect, is made to consent in advance to a range of potential changes within which the government subsequently may unilaterally order changes. Although the contractor is provided with certain rights in the face of a change order, such as the right to additional compensation for changed performance, he is generally under an obligation, unless the change order exceeds the scope of the "contracting officer's contractual power unilaterally to make changes in the contract . . . , to continue performance despite changes in the work or disputes over the amount of compensation. . . ."[32]

In addition government contracts often include one or two types of termination clauses, both of which tend to favor the government. One such clause, the default clause, spells out rules for termination by the government for cause, and the other, the "termination-for-convenience-of-the-government" clause, provides for government termination without cause. The latter clause, the more interesting of the two, exculpates the government from comprehensive common law liability for breach of contract in terminating a contract not first violated by the other side.[33]

Another aspect of government regulation by contract is provided by standard terms which require contractors to exhaust administrative remedies before they are permitted to bring claims in court. The complexity and partiality of such a requirement can be seen in the settlement machinery established by the "termination-for-convenience-of-the-government" clauses:

> the clause provides that termination claims shall be settled by agreement between the contractor and the contracting officer or, in the absence of such agreement, by an ex parte determination of the contracting officer [who thus acts as administrator and trial judge in the same case,] in

accordance with the formula stated in the clause. In either case the clause provides for administrative review by a settlement review board. Settlement review boards [, however,] are not intended to function as independent, . . . [quasi-] judicial bodies, giving approval or disapproval after impartial assessments of settlements. Their prime function is "to determine the overall reasonableness . . . *from the standpoint of protecting the government's interest."* They provide, then, an administrative check. . .

However, an independent . . . [quasi-judicial organ within the executive also can be called into play] under the clause. By cross reference to another contract provision, the clause gives the contractor the right to appeal ex parte settlements to the contract appellate agency. . .[34]

Notwithstanding such provision for a quasi-judicial, ideally independent review under the disputes clauses, however,

it can be assumed that a number of practical, recurring factors contribute to the "non-use" of disputes procedure. These may range from the burden of delay and expense through lack of confidence in the administrative tribunal to a fear, often well-founded, that appeals and protests may produce bad working relations with the contracting agency and ultimately affect the award or administration of future contracts.

What conclusion can be drawn from this analysis? Simply that the executive officers who draft contract clauses, promulgate regulations, adjudicate disputes, and administer contract performance exercise great power with relatively few formal, comprehensive, legal controls.[35]

Moreover, in connection with government contracts, a contractor's judicial rights and the role of the judiciary—an institution so intimately tied in American history and legend to the development of free contracts—have been limited. In the first place, since the government, as sovereign, traditionally has been immune from suit on its contracts, except to the extent otherwise provided by Congress, a contractor's right to sue in court is delimited by federal legislation. Thus, as the law stands, a contractor is not entitled in a contract suit against the government to the usual equitable remedies, such as injunctions and specific performance orders, nor is it clear the extent to which a contractor-defendant can counterclaim against the government-plaintiff in suits in United States district courts.[36]

In addition, the judicial role in practice seems more broadly limited. For example, "the Court of Claims has virtually no power before a contract is awarded. After award, its role is increasingly that of *reviewing administrative records on appeals by contractors which have previously exhausted administrative remedies. These appeals are relatively infrequent* and the Court, of course, has no control over what issues are presented for decision."[37]

Further examples of the "uncontractual" nature of the relationship produced by government contracts, which suggests the function of such contracts in what in fact is an administrative system, include: (1) the subjecting of certain classes of contractors to examination to determine whether, through their contracts, they have realized "excessive profits," in which event their contracts will be "renegotiated" after the fact, to eliminate such profits—so far have we come from bargain and the free market; (2) governmental "debarment," a formal "administrative" declaration that because of violation of certain contractual clauses or statutory duties, a contractor is ineligible to receive any government contract awards for a set period; and (3) the inapplicability to government of normal common law concepts, such as estoppel and apparent authority, so as to strictly limit the capacity of government officers (agents) to bind the government on contracts not technically within their authority or in violation of internal procedural regulations.[38]

Government contracting and government regulation, as part of an administered economy, have resulted in the increasing significance of a government "contract system"[39]—the same term and administrative vision of "contracts" as applied by the Chinese (*ho-t'ung chih-tu*). Three analytically separate sub-patterns have contributed to these developments in the United States:

The first . . . is the growth in power and influence of many private organizations which is occurring as a natural outgrowth of our maturing technical-economic system. As a result, these organizations achieve a greater capacity to affect the public welfare. Their affairs become "vested with the public interest"; they enter at least in part into the public domain; they become subject to increasing regulation and control.

The contract system enters this picture in several ways. By helping to build the capacity of particular organizations, it adds to their power and influence. By giving them public functions to perform, it enlarges their identification with the "public interest." By increasing their dependence upon government, it opens the way to more effective governmental intervention in their affairs. . .

The second . . . is the delegation of public authority to private agencies. . . The delegation is not formal, but accrues informally from the performance of functions which are delegated. . .

The efforts of private organizations to identify themselves with the public good and demonstrate their public responsibility is the third [such sub-pattern] . . . It emerges in part as the response of private managers to the other two. It also reflects their search for a new legitimacy for their endeavors to replace the one which was once provided by the "rights" of ownership. . .

Taken together, these three [sub-] patterns of development are moving us rapidly toward a situation where the activities of all large-scale organizations will appear irrevocably tied to the "public interest." The trend seems a natural outgrowth of fundamental changes in our total society. . . [of which the spectacular growth of] government contracting is an important part . . . , but . . . only a part.[40]

What we have increasingly in the United States, then, is an administered market, a market in which we demand of "a giant industrialism . . . not only the service of private ends but the fulfillment of public needs."[41] In achieving this demand we have recognized that the market place is grossly inadequate as a device for formulating public needs and allocating resources in the best interest of the society.

Nowhere is the reality of contract as administrative device more evident than in those industries and firms which have become almost totally dependent on government for the sale of their products. Here, most notably in the defense and space-related enterprises, the market system simply does not exist in any usual sense. *In its place stands the contract system, reducing all significant economic decisions to administrative determination.*

The contract system is a frontier development in our economy, yet it is important to remember that many of the most vital issues in contract negotiation and administration are generically the same as those encountered elsewhere in "the great economy." The problem of setting a proper fee in a contract is basically that of setting passenger fares for the airlines or the support prices on farm commodities. The control of a contractor's personnel practices is akin to the general regulation of relationships between employers and employees through fair employment, social security, workman's compensation, and other programs. . . In each case, it is necessary to determine by an administrative judgment of the public interest those things which were once decided by the operation of the market.

This can be seen more clearly in the [government] contract situation because it appears there in purer form. *When it is in full flower, the contract system actually becomes a form of "government by administration,"* with the legislative, executive, and judicial powers applicable to an industry or a firm *all exercised through the same administrative system* and often through the same administrators.[42]

The tendency toward convergence with socialist economies that is implied above with respect to contracts is exemplified by a recent, highly controversial Court of Claims decision, which sounds as if it were taken from Soviet or Chinese practice. In this case, *Christian & Associates v. United States,*[43] the court "held that a termination-for-convenience clause contained in government administrative regulations (which the court said had the force and effect of law) [but which had not been included in the contract] was imported into . . . [the] contract even though the contractor had refused to accept it on the basis of consent and bargain."[44]

In discussing the implications of the case, a number of issues generally associated exclusively with contracts in socialist planned economies are raised by experts on federal contracting:

1) to what extent . . . are contractors deprived of their expectations when clauses to which they did not agree are found to apply to their contracts long after the time of making and probably after performance has been concluded. It does not seem . . . to be enough to say . . . that the contractor knows or should know of the existence of the regulatory requirement that a clause be used. He should be able to rely on his bargain as made. . . 3) If a contract is to be deemed to include such required clauses, what status is to be given to post-contract modifications of a clause? . . . But the most striking question of all . . . is, just what should be the relationship of regulations of the executive branch to the citizen? [45]

Another authority has said of the same case:

[The judge in the Christian case] may by his decision revolutionize Government contracting. Why have any contract at all? Shall we just write a contractor and say, we give you the job pursuant to all the appropriate regulations and in accordance with the way the Government does business? [46]

And, sarcastically, the first authors write that " 'We can wait until litigation to determine what the terms of your contract are.' All of this seems a long way from the concept of bargain and it seems very close to a tolerance of a degree of executive government not contemplated by the Constitution."[47]

Day by day we seem to approach closer to the highly standardized, relatively closely-controlled administered economy heretofore thought typical only of advanced socialist economies. Considerations of national policy increasingly dominate our economy, and day by day the role of contracts seems less in line with visions of liberal ideology.

Private "Administration by Contract"

As intimated earlier, "administration by contract" is not limited in the United States to the realm of government contracts. The combination of unequal bargaining power and standardization in the private sphere has operated in a number of important businesses to encourage vertical integration by contract, a pattern of business administration referred to in Chapter IV in connection with China's management of her economy through plans and contracts.

Vertical integration is "the functional co-ordination of one or more units in each of several stages of production [and/or distribution], so that they are all operated as a single, unified industrial process."[48] Vertical integration is aimed at cost reduction and greater stability through extended coordination.

Integration may be "backward" to raw materials sources or "forward" to distribution systems. Integration may be achieved through direct ownership of various integrated units, through contractual arrangements, or frequently through a mix of the two. Ownership maximizes control, which as in the case of China's socialized industry, for example, can be exercised at least in part and with some degree of effectiveness, by administrative directives. But the "correlatives of control are responsibility and commitment." Ownership, for example, requires self-investment. By contrast, contract offers a more flexible form of integration: "Contractual arrangements aiming at vertical integration extend over a wide range, verging at one extreme on the permanence and control of ownership integration, and at the other approaching the impermanence and lack of control of the spot-market transaction."[49] Such types of agreements as requirement, output, exclusive dealing, franchise, consignment, and agency contracts all can be utilized to achieve vertical integration by contract.

Control can be exercised in the contractual mode of integration by the threat of legal sanctions for breach of contract, by the threat of nonrenewal of the relationship (an extremely potent threat where the parties, as in the case of American automobile manufacturers and dealers, for example, are in highly unequal bargaining positions and the weaker party has made extensive investment in reliance upon the continuing quality of the particular relationship), and by other indirect or informal sanctions. The selected mix of reliance on ownership, contract, and the spot market will vary widely from business to business and depend upon general economic conditions, such as supply and demand, and relevant rates of technological and stylistic change. At the margin, few or no organizations try to administer all aspects of their production and distribution processes through vertical integration, which is to say only that they do not attempt to achieve a total "self" sufficiency through such integration, but continue to rely on unintegrated units for some of their needs (for example, even GE probably does not itself, through one of its subsidiaries, or through its carefully regulated contract system, try to intricately integrate into its operations the purchase of such rarely and sporadically used items as foreign maps).

The history of automobile dealerships provides an excellent case study of vertical integration through contract. The dealer franchise contract, largely a standardized agreement drafted by lawyers representing the manufacturer and "accepted" or adhered to as written by the dealer, has been administered essentially by the bureaucracy of the dominant party—the auto manufacturer. The courts, in the name of "freedom of contract" generally left the dealers without adequate legal redress on an otherwise enforceable contract or else held that the agreement itself was not enforceable by either party because the manufacturer did not bind himself in the agreement, which therefore was "lacking in mutuality." The practical result in both cases was to leave the dealer, the weaker party, at the mercy of the manufacturers' very substantial extra-legal sanctions for noncompliance with the manufacturers' demands. At least until the mid-1950s, "the unfairness of the dealers' position at the bottom of the manufacturers' administrative pyramids" was patent.[50] The franchise agreement, without creating substantial legal obligations on the part of the manufacturer to the dealer, effectively codified the dominant expectations and demands of the manufacturer and the sorts of formal but private procedures that would be employed in settling disputes short of resort to the courts, etc. Over the years, franchise agreements have required a dealer to take a minimum number of autos, to use his best efforts to develop his territory to the satisfaction of the manufacturer, to invest at least a minimal amount of capital as determined by the manufacturer, to adopt specified accounting and reporting systems (shades of China), to require the dealer to deal exclusively in the products of the franchiser, and so on.

Until the late thirties, when courts began to hold the agreements enforceable, the manufacturer ultimately sanctioned his demands by the threat of exercising his power to terminate the legally unenforceable franchise agreement:

> A terminated dealer frequently is unable to get a franchise from another manufacturer. Since his capital investment is so specialized that it cannot easily be transferred to other kinds of business, termination has often been called an economic death sentence. . . After a dealer has committed his capital to a franchise, protection of the manufacturer's interest does not depend on the availability of legal sanctions. In fact, legal invalidity of the franchise, precluding court control, adds to the strength of the manufacturer's non-legal sanctions.[51]

Once the enforceability of the franchise agreements was accepted, the central issue became abuse of termination clauses. As a general rule, dealers continued to lose in court as they have out of court, as manufacturers molded contractual clauses in response to judicial and legislative restraints in order to maintain their ultimate sanction in the form of clauses providing for termination at will, for cause, or clauses which required periodic renewal of the franchise if it was to continue. Courts were particularly loathe to police the manufacturer's "cancellation" of the relationship by a decision not to renew the franchise, since such supervision seemed to judges in violation of the still cherished doctrine of freedom of contract.

The relative-success story of the history of political lobbying by auto dealers' associations and the resulting reforms shows a pattern of further standardizing of the relationship. A kind of due process, for example, has been imported into many of the manufacturers' administrative decisions that most directly affect the rights of particular dealers. Similarly, the courts and legislatures by recently requiring that manufacturers exercise "good faith" in cases of termination and non-renewal, have tended to treat the creation and maintenance of the franchise in terms of "a kind of compulsory contract," which requires that "manufacturers must continue relationships with dealers unless a dealer has in some material way violated a duty under the franchise."[52] Thus, to the extent that manufacturers' one-sided vertical control has been limited, here again protection has been afforded the weaker party by transforming what was in law a contractual relationship but in fact a status-like relationship into a status-like relationship at law with reciprocal rights and obligations. Manufacturers' "administration by contract" has been restricted by government pressure and regulation.

The Role of Contracts in Business: China and the U.S.

The case of the automobile industry, in which for approximately thirty years franchise agreements were not recognized as enforceable contracts, raises an even broader issue regarding the role of business contracts in America. This issue involves the relative roles of nonlegal, socioeconomic relations and of contracts in business relationships. The issue appears to have been given only sporadic attention over the last few generations, and it is only relatively recently that serious empirical studies have been undertaken of conditions in the United States.[53]

The suggestion that nonlegal, informal institutions may perform the critical roles in exchange transactions in the business community of our advanced industrial society seems to cut against the grain. As indicated earlier, it has been common, and within limits useful, to understand societal development or advance substantially in terms of increasing "rationalization." When we speak of development, one of the core phenomena we usually refer to is the increased use and refinement of rational, purposive organizational techniques, one such being the institution of contracts. This means by implication that: (1) inevitably as an underdeveloped country like China develops, it will come increasingly to rely on formally structured commercial exchange relationships, like contractual exchanges; and (2) the characteristic mode of commercial exchange relations in an advanced economy is contractual. Development, in short, implies increasing reliance on contracts.

Two factors may account for this one-sided understanding of development and of the role of formal-legal institutions in development. The very growth of legal-rational elements before the eyes of recent historical consciousness may have led us to underplay the crucially important other side of the development pattern, the continuing vitality of informal, interpersonal, flexible elements in developing and in advanced industrial societies. The impact of "rationalization" in history, perhaps, has been enhanced further by a misunderstanding of Max Weber's influential conception of development. In categorizing ideal types of authority and implying that the legal-rational type is most advanced, Weber did not mean to suggest that societal realization of the legal-rational ideal type was a possible end state of development. Weber believed that development or modernization implied greater rationalization, but he did not foresee a purely rational society.[54] Nevertheless, at least until relatively recently, theorists, seeing Weber's emphasis, have tended to think of development in terms of the elimination of traditional, irrational, ascriptive, personal elements in the society. They assumed that legal-rational forms increase strikingly during development, and that in advanced industrial society such forms are highly dominant.

This predisposition has been reinforced in the field of legal education by the case method of studying law. If one were to rely exclusively upon American court decisions as empirical data for characterizing the area of commercial exchange relationships, one undoubtedly would conclude that contracts are the characteristic, if not the exclusive, mode of structuring such relations. This orientation has natural appeal to lawyers. One does not have to be a subtle practitioner of the sociology of professions to understand that lawyers would, and in fact do, tend to emphasize the formal-legal side of structured exchange relationships. Lawyers, including myself, have tended to focus on the need for elaborate contractual rules, the purpose of which is to secure the expectations of the parties to the contract while simultaneously minimally protecting society's interests.

The lawyers' view is always in danger of becoming formalistic, of stressing rule-following without taking account of broader, nonlegal functional considerations. At its best it will try to take account both of such functional considerations and rule following. The rather abstract legal-functional analysis of Chinese contract law in Chapter III is an attempt to explore how Chinese contract law, as expressed in a theoretical work, might well prove to be functional given the societal context and the government's priorities.

Judging by the limited evidence we have of Chinese practice, at least in some areas it seems to conform broadly to theory. My impression, for example, is that the ideal of specific performance as the normal remedy is substantially reflected in practice. However, when one moves on to consider practice with regard to other Chinese contract rules that are as readily justifiable from a legal-functional approach, one finds, with acute embarrassment, that practice appears to be "dysfunctional." Rules requiring the assessment of penalties for breach of contract, for example, while explicable on legal-functional grounds as necessary to encourage specific performance, in fact are not systematically enforced. In practice, remedies appear to be applied on a much more broadly pragmatic, often highly discretionary, basis than one might gather on the basis of purely legal-functional considerations.

This "distressing" fact prompts two related thoughts about the legal-functional approach. First, the legal system in China, and anywhere, is not as rationalized as a pure exercise in legal-functional terms presupposes. Second, one cannot understand even legal practice solely in terms of legal functionalism, if only because other nonlegal factors always impinge upon the legal system in practice. Insofar as functionalism is not taken as merely a tautology, then, this suggests that few, if any, disciplines can be sufficiently isolated or rationalized to be explicable purely in their own functional terms.

What I am suggesting is that we must begin to construct typologies of formal and informal modes of structuring exchange relations available in various societies at selected times. For only by understanding the interaction between these modes can we ascertain the relative significance, distinctive quality, and utility of each in practice.

The importance and distinctiveness of the informal (or nonlegal) mode in the United States, for example, was suggested some time ago by Karl Llewellyn, when he wrote of legal and non-legal obligations:

> The law always reaches for and commonly approximates some single definite manner and quantity of performance as the measure of the obligation. The measure is to be fixed and inflexible from the time of contracting. . . [On the other hand,] non-legal obligation is in essence flexible within wide limits, and remains flexible even after the . . . [contract] has been made. It runs not in a line without width, but in a belt or a range of permissible performances. What is due, what can be demanded, what can be offered, is anything within the belt limits. . . So that, pending performance, either party can take the initiative in fixing performance more narrowly. The seller, by tendering a delivery within the permissible limits (or by demanding a permissible alteration). The buyer, by specifying what he needs, precisely (or by demanding a permissible alteration). Before such narrowing, the range of proper performance is much wider than is commonly recognized at law. Even after such narrowing, say by the buyer, it may still be wider: for even timely specification leaves the seller such leeway as practical reason requires for a seller in the circumstances. . . [W]hat we have been accustomed to think of as habit, or custom, or folk way, or practice, or institution is not a line-concept, but a belt concept, with an important range of variation. . . [T]here are two ranges in each case: an inside range of permissible (unobjectionable, unnoticed, not-adversely-reacted-to) variation, and an outside range of nonpermissible (objectionable, noticed, adversely-reacted-to) variation.[55]

As numerous commentators have pointed out and as businessmen soon learn, contract rights and remedies in reality are neither an adequate substitute for performance nor an adequate sanction for failure to perform.[56] While, at a minimum, "legal enforceability is sometimes a factor in inducing performance by a debtor . . . [and] a salvage factor it always is,"[57] the non-legal sanctions and incentives—in combination with such structural facts of business life as the organization of enterprises for the very purpose of performing their obligations in the ordinary course of business—[58]play a crucial role in stimulating performance of one's commitments.

A preliminary study of the business practices of selected Wisconsin manufacturing firms, for example, illuminates the interpenetration of legal and nonlegal modes and suggests that, at least in highly institutionalized relationships, formal contracts even in the most advanced industrial society in the world may not be the dominant form of structuring exchange relations.[59] Employing a sociological rather than a legalistic approach to the institution of contracts, the study treats contracts as devices for conducting exchanges. The term "contracts" is taken itself to constitute a spectrum of different sorts of relationships, which then may be described as more or less contractual.

In rejecting the dichotomous legal approach of contract/no-contract, this study, carried out by Stewart Macaulay focuses on two elements of contracts: (1) the degree of rational planning, with careful provision for future contingencies, involved in the creation of the relationship; and (2) the orientation toward legal sanctions to induce performance or sanction breach. With regard to the first, Macaulay found that "a significant amount of creating business exchanges is done on a fairly noncontractual basis. . . Many, *if not most*, exchanges reflect no planning, or only a minimal amount of it, especially concerning legal sanctions and the effect of defective performance.[60]

More specifically with regard to (2), Macaulay found the firms studied were relatively unconcerned with planning their transactions so they would be legally enforceable. Thus, although it was widely recognized that requirements contracts (X promises to supply all of Y's needs of certain goods during a certain time period and Y "promises" to accept them) probably are unenforceable in Wisconsin, such contracts were rather commonly concluded. Macaulay believes that even though a substantial amount of creating commercial exchanges is structured in a fairly noncontractual manner, still "the creation of exchanges usually is far *more* contractual than the adjustment of such relationships and the settlement of disputes."[61] In short, to generalize from this study, legal-rational, formal, contractual modes of structuring commercial exchange relationships appear to be only one, and perhaps not the dominant, significant mode even in the United States.

This conclusion implies that at least a substantial proportion of commercial exchange relations should be viewed in terms of *ongoing process*, rather than in terms of predefined, relatively inflexible rights and duties. These exchange relations, then, are in flux, frequently manifesting a series of readjustments mutually acceptable to the parties because of the benefits of the continuing relationship. Insofar as the reality of exchange relationships (or, for that matter, of planning) is process rather than highly articulated formal structure, a highly legalistic approach to the reality gives it an artificially static quality.

The legalistic approach, moreover, seems peculiarly misrepresentative of how the parties themselves see the relationship. Even where the parties in Macaulay's study had detailed and carefully planned agreements, there was a "hesitancy to speak of legal rights": "'You don't read legalistic contract clauses at each other if you ever want to do business again.'" Lawyers were said not to understand "the give and take needed in business."[62] And, in fact, law suits for breach of business contract were rare.

The institution of contracts is only one of a variety of modes for structuring social relationships, one purpose of which is to ensure predictability and security in business exchanges. Therefore, when other functionally equivalent modes are available, the formal-legal mode may be ignored.[63] To a substantial degree, then, one must understand the sociological context of the exchange to understand how and why the formal legal institutions will be called into play. For example, given the relative absence of continuing business relationships, parties to one-shot transactions, like the sale of the New York Yankees, are likely to rely heavily on contractual formality. Or again, parties to a highly speculative transaction, even if a significant continuing relationship is present, may rely extensively on legal formality in order to allocate the various risks in advance, especially if the stakes in the transaction are thought to be disproportionately high relative to the continuing relationship. In such cases the nonlegal relationship alone might not serve as an effective sanction.

Macaulay's findings, suggesting the importance of noncontractual elements in exchange relations in the United States, indicate that Malinowski's earlier emphasis on reciprocity as the crucial element in primitive societies for informally structuring continuing exchange relations may be more broadly relevant to highly developed societies than previously imagined.[64] One might say that new "particularisms" (status-like) evolve in more advanced societies to replace traditional particularisms. While it may be true that the new particularisms of continuing, interpersonal business relationships evolved originally from highly specific, legal-rational business relationships, the end product of the new particularism broadly parallels traditional, diffuse, interpersonal relationships. The significance for us of such continuing relationships is that such relationships may act as their own informal sanctions. The fact that the parties to the relationship expect the exchange relationship to continue and to be mutually beneficial in the future substantially determines their conduct in the relationship. In such a context, they tend not to rely on legal forms or sanctions.

To a substantial degree the informal continuing relationship embedded in a business subculture performs two of the major functions of contracts. First, as suggested above, the relationship's informal, diffuse incentives and sanctions frequently are more effective than formal-legal ones. Second, the understanding of the other party's needs and priorities that each of the parties gains by relating to the other over time may be greater than that gained by the more explicit communication of such information through more formal negotiation and dispute settlement within the contractual context.

Given these perspectives, in what contexts and/or phases of societal development are contracts likely to be the characteristic and dominant mode of ordering commercial exchange relations? As tentatively suggested above, contracts, like formal legal institutions in general, are likely to be relied upon most heavily in times and places where informal trust relations of sufficient scope or relevance do not exist. One might expect, then, that in a period of rapid economic and social change, such as that of the English Industrial Revolution, there would be a strong tendency to rely on contractual devices. As societies stabilize themselves, even when as in advanced industrial societies "stabilization" means the institutionalization of change, I would hypothesize that commercial business relations would tend to rely less on formal-legal institutions for structuring recurrent transactions. In stable societies, where change is assimilated by pre-existing relatively diffuse groupings and relationships, these groupings will provide the most meaningful framework for many exchanges.

If all this is true, then a definitive evaluation of the institution of contracts in China depends upon an understanding of the broader, socioeconomic context in which "businessmen" in China find themselves. Unhappily, almost nothing is known in depth of business practice in China. With the possible exceptions of the work of Barry Richman, Franz Schurmann, and Audrey Donnithorne, none of whom has really scratched the surface in this regard, hardly any scholar has even addressed himself to such issues.[65]

One is left, then, with speculation. One might speculate, on the one hand, that to the extent commercial business relations in China remain based in large part on historical relations, or to the extent that new, relatively stable, continuing relations have evolved since 1949, the attempt to impose strict contract disciplines on business relations, at best, would be superfluous, or at worst, at odds with the strong and perhaps pragmatic proclivities of the relevant parties. Continuing with this line of reasoning, on the other hand, to the extent that the Chinese Communist revolution—the repeated campaigns and planning—have effectively shattered prior business relationships and pre-vented new ones from emerging (a very difficult feat, one would imagine, as evidenced by the targets of the assaults embodied in the Cultural Revolution), reliance on more formal, legal modes of structuring might well have become more essential.

Chapter VI

THE MEANING OF CONTRACTS: THE QUESTION POSED

The institution of contracts is a somewhat problematic one in at least three related aspects. (1) As a universal concept, it is not entirely clear what remains distinctive in the nature of contracts. (2) In the planned sector of Communist China's economy, the functions performed by contracts arguably appear so minimal as to raise doubts whether they are significant or distinctive. And (3) contrary to common assumptions, societal development does not seem to result in a unilinearly increasing reliance across the board upon the institution of contracts. While the absolute number of contracts concluded seems to increase, the significance of contracts from area to area within the society does not reflect a unilinear progression. So, the role of contracts over time in development also is problematical.

(1) To some extent the answer one gives to the question what is distinctive in the nature of contracts depends upon the perspective with which one asks the question. A perspective earlier employed is one that distinguishes contracts as an institution both from ideal (or pure) types of authoritative commands and wholly voluntary agreements. From this perspective commands are seen as emanating wholly from above, as involving no element of agreement. The polar opposite of commands are wholly voluntary agreements in which the parties determine all the procedures and substantive terms by agreement, without interference at any point by authoritative governmental institutions. In this perspective, contracts are in the middle, somewhere between pure commands and purely voluntary private agreements.

One trouble with this perspective, however, is that, despite its heuristic value, it is in some measure inappropriate to compare the actual institution of contracts with an ideal type of commands. What this perspective obscures is that commands (here read also as plans) in practice also involve elements of agreement, of at least tacit reciprocity. Commands, if they are to function with minimal efficiency over time and on a society-wide basis, must incorporate elements of agreement. For authority to be accepted over time, it must reflect to some degree the wishes and capacities of those over whom it is being exercised. Thus, in practice, commands (plans) and contracts in all societies fall somewhere between the polar extremes of pure agreement and pure command. If there is something "distinctive" about contracts, then, it is that, as a complex institution, it provides facilities for mixing the same elements of authority and agreement embodied in commands in an amalgam that always incorporates a greater degree of agreement than do commands.

Given this second perspective, the degree of difference between plans and contracts can be narrowed to the point where they become nearly indistinguishable. If, for example, the authoritative element in contracts is strengthened, if the state imposes a duty upon the parties to contract, if the scope of the agreement (bargain) is authoritatively limited, and if the contract's integrity is undermined, then it is the fundamental similarity between contract and plan, rather than the fundamental difference between them, which becomes manifest. Plans and contracts, then, can be seen as two sides of the same process of defining future expectations and behavior.

(2) Economic contracts within the planned sector in Communist China are part of the planning process. Plans are formulated and implemented upon a foundation of thousands of pieces of paper, some of which are contracts. But, given the context, do contracts perform significant and distinctive functions? I have no certain answer.

It is clear that contracts do perform a function of detailing plans. And, of course, outside the strictly planned areas, where the scope of the contractual bargain is broader, contracts may facilitate the same sort of "private law-making" or decentralized decision-making associated with contracts in the West. Still, within the planned area, this role of adding greater specificity to plans at the production unit level would hardly seem to justify the high praise the Chinese consistently have lavished on contracts. What other functions do or might contracts perform in the planned sector of the economy?

First, contracts are said to inculcate a feeling of responsibility on the part of the parties who "accept" the terms of the contract. But in fact contracts seem likely to perform this function only to the extent that the terms they embody are the product of voluntary agreement. To the extent the terms are more in the nature of commands "imposed" from above, as appears frequently to be the case in Communist China, the desired psychological response to signing a contract is less likely to occur and, if it does occur, might well have been elicited by commands in the first place.

But perhaps even in the most highly planned areas, the contractual elements of party participation and of solemnization should not be dismissed so lightly. A decision in the making of which one has participated to even a minimal degree is likely to appear more just to the participant than one in which he did not participate at all. When one agrees to certain terms, whatever the pressures driving one to consent (short of actual physical duress), it will seem more just that one should perform as "agreed" than if one had never so agreed.

Similarly, the quality of justness will also be enhanced from the dominant party's perspective. We too often forget that rulers, political leaders, monopolists, and captains of industry also have concepts of justice which in recent centuries undoubtedly require that those whom they dominate in fact give some sign of their consent to the relationship. In this sense, whatever standardized terms may be authoritatively imposed by government upon a relationship, contract retains some remnants of its nineteenth-century pure type, that justice is what the parties have said it is. The parties' expectations as laid out in the contract, whatever their expectations might have been prior to the making of the contract, influence their vision of what is due them. The theory that societies are based upon social contract, while historically apocryphal, reflects an awareness of this fact and a recognition of the consensual element in all authority relations.

Second, contracts are said to provide a horizontal bond between legal equals, to which civil law obligations and remedies attach and which therefore stimulates decentralized supervision and enforcement procedures essentially based on party initiative. By contrast, plan relationships are thought of as essentially vertical relationships in which the primary burden of supervision and enforcement falls upon administrative agencies superior to the relevant production unit. In this view, contracts provide an alternative and important channel for stimulating performance of obligations necessary for plan realization. Civil law rights and remedies are seen as incentives to the parties to signal deviations from the more-or-less planned contracts. Viewed in a broad perspective, then, contracts are a facility for the delegation of plan administration to concerned production (or distribution) units, which will be judged on the basis of their contract and plan fulfillment.

There are, however, a number of objections to this reasoning as well. In the first place, as recent developments in the Soviet Union appear to suggest, there is no inherent reason why civil law obligations could not attach to plans and planning acts. Except for the admitted quasi-administrative function of providing the detailization necessary both to implement plans and to create enforceable civil law obligations, contracts then might lose most of their functions. This, in fact, appears to have happened in the Soviet Union within the limited realm of very strictly planned transactions. In this area plans and planning acts (that is, administrative acts) are very detailed. Here, contracts hardly appear to perform even a detailization function and civil law obligations arise from plans and planning acts alone. Such planning acts, moreover, relate production units horizontally to each other, as well as vertically to the administrative superstructure. In this area contracts truly appear to be superfluous. Nevertheless, it is at least true that it would be a mammoth, perhaps impossible, administrative undertaking to try to administer the entire planned economy in as detailed a manner as would be necessary to make contracts wholly superfluous.

But if one must grant that contracts are useful in decentralizing the detailing and assigning of planned tasks down to firms within the planned

economy, it remains, however, a somewhat separate question whether, beyond
that, contracts are an added stimulus to performance of those planned tasks.
To what extent, for example, do parties enforce contractual terms against one
another? One would expect, given the incentives provided in the institution
of contracts in the Chinese and Soviet economies, that the answer would be
often and substantially. This expectation seems to be borne out by estimates
of the number of cases handled by Soviet State and Departmental Arbitrazh
organs annually, about 1,000,000 in total, which constitutes about .4 per cent
of the estimated number of planned contracts concluded each year.[1] To what
extent these cases are litigated at the initiative of Arbitrazh rather than the
parties, however, and are confined to pre-contract, predominantly administrative, disputes is not clear. It may well be that the number of party-initiated
suits for breach of an already concluded contract is substantially lower than
the total.

It appears that in the Soviet Union, Hungary, and China, parties in fact
are very reluctant to enforce contract rights and to maintain strict contract
discipline: "In the field of contract performance the tendency to 'mutual
amnesty' . . .—the failure to claim sanctions for improper performance of
contract—. . . seems to be widespread."[2] Despite the impetus the planning
structure gives to concluding and performing contracts, there appears to be
significant hesitation on the part of contracting parties to enforce their legal
rights. The reason most often given to explain this fact is that a prospective
plaintiff who "depends upon the output and future deliveries" of the defaulting party, "might be willing to sacrifice his formal right to a sanction for
the goodwill of his partner." While commentators tend to limit their recognition of this practice of "mutual amnesty" to the refusal to enforce penalty
clauses, more general considerations about the role of the formal-legal institution of contracts in continuing relations might be taken to suggest that "mutual amnesty" may well be an even more widespread phenomenon than previously had been imagined. As suggested, it appears that in light of these
more general considerations the duty of a party to a planned contract to enforce his contractual rights in the case of his promisor's default frequently
may fly in the face of demands for flexibility and tolerance inherent in a continuing relationship between business enterprises. In this sense the stimulus
to strict performance of contracts provided by the planned context may be
said to be disruptive to normal business relations, and, perhaps, thereby to
production itself.

(3) If the suggestion that requiring strict performance of contractual
obligations may be injurious to normal business relations still sounds strange,
it may be because we continue to hold to a one-sidedly legalistic orientation
toward modern commercial exchange relations. When we think of institutions

for providing stability of expectations, for enhancing calculability, and so forth, almost by second nature we think of the modern legal system. By contrast, reliance upon personal relationships—taken to be characteristic of traditional societies in general, and of much of Chinese history in particular— seems to mean irrationality, instability, and insecurity. But here we have been victimized by our own original insight and ideology to the point that we have lost sight of the complexity of reality.

NOTES

Abbreviations Used in the Notes

CB   *Current Background.* United States Consulate General, Hong Kong.

CNA   *China News Analysis*

ECMM *Extracts from China Mainland Magazines.* United States Consulate General, Hong Kong.

JMJP   *Jen-min jih-pao* (People's daily)

JPRS   Joint Publications Research Service. United States Department of Commerce, Office of Technical Services.

NCNA New China News Agency

NFJP   *Nan-fang jih-pao* (Southern daily)

SCMP *Survey of China Mainland Press.* United States Consulate General, Hong Kong.

TKP   *Ta kung pao* (Impartial daily)

Introduction

1. See, for example, Philip Green, "Science, Government, and the Case of RAND: A Singular Pluralism," *World Politics* 20:301–326 (January 1968).

2. For an article that analyzes such problems in connection with comparative criminal law systems, see my "Crime and Punishment: China and the United States," *World Politics* 21:152–181 (October 1968).

3. Edwin W. Patterson, "Compulsory Contrasts in the Crystal Ball," *Columbia Law Review* 43:742 (July 1943).

I. *Contracts in the Anglo-American Tradition*

1. For excellent discussions of a number of these issues, see generally, Karl N. Llewellyn, "What Price Contract?—An Essay in Perspective," *Yale Law Journal* 40:704–751 (1931), esp. at 704–708; and Morris R. Cohen, "The Basis of Contract," *Harvard Law Review* 46:553–592 (February 1933).

2. Roscoe Pound, "The New Feudalism," *American Bar Association Journal* 16:553–558 (1930).

3. Lawrence M. Friedman, *Contract Law in America* (Madison, 1965), pp. 20–22 (italics added). "Consideration," Friedman writes at p. 21, "is often said to be one of the prerequisites of a legally enforceable contract. The concept in the law of contracts is complex, ubiquitous, and tricky. The core idea, as it came to be understood in the Nineteenth Century, is that a contract to be enforced ought to be an exchange, that is, a two-sided arrangement, with something contributed by both sides. In the ordinary contract, it is easy to find 'consideration' . . . A promise to make a pure gift, however, 'lacks consideration,' because it is said that the donor gets nothing in return. The classic doctrine, that the law will not look into the adequacy of consideration, means that . . . [as to an agreement for the sale of land at a price of $1000] a court will enforce the contract to convey the land and will refuse to consider whether $1000 is or is not a 'fair' or 'adequate' price."

4. Ibid., p. 17 (italics added).

5. Friedrich Kessler and Malcolm Sharp, *Contracts, Cases and Materials* (New York, 1953), p. 14.

6. Lawrence M. Friedman, pp. 18–19 (italics added).

7. Ibid., p. 18.

8. See chap. 5.

9. Llewellyn, "What Price Contract?", p. 716 (original italics).

10. Emile Durkheim, long ago, brought out one relationship between contracts and development. Development for Durkheim essentially meant increasing division of labor among component groups within society, increasing specialization of function. The family and clan, for example, tend to lose their diffuse functions; production tends increasingly to be centered outside the home. Specialization and division of functions means greater interdependence between component units, many of which previously were self-sufficient. Increasing complexity of the economy, arising from specialization and interdependence, requires more sophisticated methods of coordinating exchanges between units. It requires, for efficiency, a high level of decentralization of decision-making to the units or parties directly involved in commercial exchanges. Contracts, based upon the agreement of the parties, serve that function. And, with greater interdependence, and with greater scale, what happens to one unit of society tends to have greater social repercussions throughout the society. Therefore, rules are formulated to regulate contractual relations in the interest of society at large and in the interest of the contractual institution as well. Emile Durkheim, *The Division of Labor in Society* (New York, 1964), chap. 7, pp. 200–229.

11. Kessler and Sharp, p. 3; and see Llewellyn, "What Price Contract?"

12. Julius Stone, "The Myths of Planning and Laissez Faire: A Re-Orientation," *George Washington Law Review* 18:1–49 (December 1949), esp. at pp. 15–23.

13. Henry Sumner Maine, *Ancient Law,* Pollack's ed. (1930), p. 165.

14. Excellent discussions of these developments can be found in Roscoe Pound, "The End of Law As Developed in Juristic Thought," *Harvard Law Review* 30:201–225 (January 1917); and Pound, "Liberty of Contract," *Yale Law Journal* 18:454–487 (1909).

15. Herbert Spencer, *Justice,* section 70, as quoted in Pound, "Liberty of Contract," p. 484.

16. A remark by Lord Northington in *Vernon v. Bethell,* 2 Eden, 110, 113, as quoted in ibid., pp. 471–472. In this same vein, Karl Llewellyn has written that "'Agreement' does not even today carry any necessary connotation of real willingness. Acquiescence in the lesser evil is all that need be understood. The problem of 'reality of consent' is essentially one of determining what types of pressure or other stimuli are sufficiently out of line with our general presuppositions of dealing to open the expression of agreement to attack. 'Economic duress' is still the order of the day save for particular exceptions (usury, equity of redemption, duress of goods) . . . Truly 'agreement' is an elusive concept." Llewellyn, "What Price Contract?", p. 728.

17. Pound, "Liberty of Contract," esp. at pp. 470–487.

18. Lawrence M. Friedman, pp. 19–20.

19. Ibid., pp. 23–24; and see Nathan Isaacs, "John Marshall on Contracts: A Study in Early American Juristic Theory," *Virginia Law Review* 7:413–428 (March 1921), esp. at pp. 427–428. And for a general, but quite uneven discussion of this point and others, see Harold C. Havighurst, *The Nature of Private Contract* (Evanston, 1961).

20. Lawrence M. Friedman, p. 25.

II. *Contracts and Planning in China, 1949–1963: An Overview*

1. See generally George H. Hampsch, "Marxist Jurisprudence in the Soviet Union: A Preliminary Survey," *Notre Dame Lawyer* 35:525 (1960); and see Harold J. Berman, *Justice in Russia* (Cambridge, Mass., 1950), pp. 9–17.

2. Berman, *Justice,* p. 12.

3. Lon L. Fuller, "Pashukanis and Vyshinsky: A Study in the Development of Marxian Legal Theory," *Michigan Law Review* 47:1157–65 (June 1949).

4. Berman, *Justice,* pp. 17–50.

5. Quoted in Berman, review of *Settling Disputes in Soviet Society: The Formative Years of Legal Institutions* by John N. Hazard, *Yale Law Journal* 70:686 (March 1961).

6. Fuller, "Pashukanis and Vyshinsky."

7. Berman, "The Current Movement for Law Reform in the Soviet Union," *The American Slavic and East European Review* 15:181 (1956).

8. Ibid.; see also Berman, *Justice,* p. 33.

9. Ibid.

10. Ibid.

11. Ibid.

12. Berman, "Soviet Law Reform—Dateline Moscow 1957," *Yale Law Journal* 66:1213 (1957).

13. Quoted in Berman, "Current Movement," p. 181.

14. Fuller, "Pashukanis and Vyshinsky," p. 1163.

15. Berman, *Justice,* p. 38.

16. Ibid.

17. Berman, "Soviet Law and Government," *Modern Law Review* 21:23 (January 1958).

18. Ibid.

19. Berman, "Soviet Law Reform," pp. 1211–15.

20. Berman, "Soviet Law and Government," p. 26.

21. Berman, *Justice,* p. 35.

22. This point is expressed or implied in all of Harold J. Berman's writings referred to above in notes 1, 5, 7, 12, and 17.

23. See generally Jean Escarra, *Chinese Law* (Peking and Paris, 1936), tr. G. R. Browne, for Works Progress Administration, W.P. 2799 (University of Washington, Seattle), originally titled *Le Droit Chinois: Conception et Evolution, Institutions Legislatives et Judiciaires, Science et Enseignement,* pp. 97–119.

24. See, for example, the statements extolling the virtues of "socialist civil law" in *Chung-hua jen-min kung-ho-kuo min-fa chi-pen wen-t'i,* written by the Institute of Civil Law, Central Political-Legal Cadres' School (Peking, 1958), tr. as *Basic Problems in the Civil Law of the People's Republic of China,* by JPRS, 4879, CSO: 1473–5 (1961), pp. 22–23.

25. Ibid., p. 27.

26. See generally T. J. Hughes and D. E. T. Luard, *The Economic Development of Communist China, 1949–1958* (London, 1959), chap. 1; R. H. Tawney, *Land and Labor in China* (New York, 1932); for a brief sketch of the overall economic development of the People's Republic and the problems it has encountered, see Li Choh-ming, "The First Decade, Part II, Economic Development," *The China Quarterly,* No. 1: 35–50 (January-March 1960).

27. E. F. Szczepanik, "The Principles of the Economic Policy of Maoism," *Contemporary China,* I, 50 (Hong Kong, 1955).

28. Ibid.

29. Hughes and Luard, chap. 2.

30. The Common Program was passed by the Chinese People's Political Consultative Conference on September 29, 1949. It appears in translation in *CB,* no. 9:1–14 (Sept. 21, 1950).

31. The Chinese People's Political Consultative Conference represented the "united front" between the Chinese Communist Party (CCP) and the other "democratic" parties.

32. *CB,* no. 9:7–10.

33. Ibid., p. 7, Article 28.

34. Ibid., p. 7, Article 29.

35. Ibid., p. 8, Article 30.

36. Ibid., p. 8, Article 33.

37. For brief summaries of the impact of the Common Program see Ronald Hsia, *Economic Planning in Communist China* (New York, 1955), pp. 7–9, 74–75; and see Chao Kuo-chun, *Economic Planning and Organization in Mainland China* (Cambridge, Mass., 1959), I, 5–6.

38. Hughes and Luard, chap. 3.

39. Ibid.

40. Ibid., chap. 7.

41. Ibid.

42. Ronald Hsia, pp. 9–16.

43.   Ibid.

44.   G. L. V. Hooton, "The Planning Structure and the Five Year Plan in China" *Contemporary China,* I, 92 (Hong Kong, 1955).

45.   Chang Yu-nan, "Industrial Administration in Communist China," *Western Political Quarterly* 9:854–855 (December 1956).

46.   The Government Affairs Council was known as the State Council after 1954 and hereinafter will be referred to as such.

47.   Chang Yu-nan, pp. 854–855.

48.   Ronald Hsia, p. 9, n. 39.

49.   Hooton, p. 93.

50.   The State Planning Committee, since 1954 called the State Planning Commission, hereinafter will be referred to as such. Prior to that time the State Planning Commission was directly under the Central People's Government Council (the analogue before 1954 of the National People's Congress); after 1954 the Commission was under the State Council. See E. S. Kirby, "The Central Structure of Economic Planning in Communist China," *Far Eastern Economic Review* 22:737, 738, 801 (June 13, 1957).

51.   Ibid.

52.   See generally with respect to the planning method the articles by Kirby and Hooton; and Chao Kuo-chun, pp. 2–5.

53.   Chang Yu-nan, pp. 861–862.

54.   Tu Hsiang-kuang, "The Business Accounting System, A Prerequisite to Socialist Industrialization," *Hsueh-hsi* (Study), no. 6 (June 1955), tr. in *SCMP*, no. 1106:35–42.

55.   See Zoltan M. Mihaly, "The Role of Civil Law Institutions in the Management of Communist Economies: The Hungarian Experience," *American Journal of Comparative Law* 8:310, 312–313 (1959).

56.   Wu Yuan-li, *Economic Survey of Communist China* (New York, 1956), pp. 202–207.

57.   Ibid.; and see Mihaly, p. 312.

58.   Li Po-jan and Chang Chao-chen, *The Theory and Practice of the Economic Accounting System,* from which excerpts on pp. 63–65 and 272 are quoted in Chao Kuo-chun, pp. 66–68.

59.   Tu Hsiang-kuang, p. 40.

60. Ibid.

61. Li and Chang in Chao Kuo-chun, p. 68.

62. Ibid.

63. See below for further discussion of this topic with regard to subsequent periods.

64. Wu Yuan-li, *Economic Survey,* pp. 220–227.

65. Chang Yu-nan, pp. 864–865; *Basic Problems,* p. 31.

66. See R. H. Tawney.

67. See above, p. 13.

68. The Constitution of the People's Republic of China was adopted on September 20, 1954, by the First National People's Congress. Its provisions appear in translation in *CB,* no. 297:1–25 (Oct. 5, 1954).

69. The important relevant features of the Constitution are summarized and briefly discussed in Chao Kuo-chun, p. 6.

70. Ronald Hsia, p. 73.

71. Peter S. H. Tang, *Communist China Today* (Washington, 1958), I, 298.

72. Hughes and Luard, chap. 4.

73. The sources of information regarding the planning structure are as follows: Kirby; Chao Kuo-chun, esp. pp. 1–45; Hooton; Chang Yu-nan; Peter Tang, chaps. 5 and 6; and Li Choh-ming, *The Statistical System of Communist China* (Berkeley and Los Angeles, 1962), pp. 22–27.

74. Chao Kuo-chun, p. xv.

75. See Peter Tang; and Li Choh-ming, *Statistical System,* pp. 22–27, for a general discussion of the government structure.

76. In fact the structure is somewhat more complicated. See Peter Tang, p. 225.

77. Ibid., pp. 225–226.

78. Kirby.

79. Ibid., pp. 738 ff.

80. Ibid., p. 739; Chao Kuo-chun, pp. 1–2.

81. Hooton, p. 96.

82. Li Fu-ch'un on State Planning at September 24th Meeting of Eighth Party Congress, New China News Agency, September 24, 1956, tr. in *SCMP*, No. 1378:3; Chao Kuo-chun, p. 1.

83. Li Choh-ming, *Statistical System*, p. 13.

84. Kirby, p. 738.

85. Li Choh-ming, *Statistical System*, p. 24.

86. Kirby, pp. 801 ff.

87. Chao Kuo-chun, p. 1.

88. Kirby, pp. 803–804.

89. Ibid., pp. 749 and 804.

90. See Wu Yuan-li, *Economic Survey*, pp. 550–551; and Li Choh-ming, *Statistical System*, p. 21.

91. Li Choh-ming, *Statistical System*, pp. 22–23.

92. *Hsin chien-she* (New construction), no. 2:12–16 (Feb. 3, 1957, tr. in Chao Kuo-chun, pp. 95–104; and see ibid., p. 55.

93. *Basic Problems*, pp. 213–214.

94. Kirby, pp. 738 and 741; Chao Kuo-chun, p. 5; Chang Yu-nan, p. 871.

95. See Chang Yu-nan, pp. 870–871.

96. Ibid., pp. 867–869; and see Kirby, p. 741.

97. Ibid.

98. *Basic Problems*, p. 31.

99. Ibid., p. 201.

100. See below with regard to such use in the next period.

101. See below, pp. 37–38, for a discussion of general, sub-, and direct contracts.

102. Yang Ch'un-hsu and Li Cheng-jui, "Apply the Contract System Extensively," *TKP* (Feb. 20, 1959), tr. in *SCMP*, no. 1980:8–29. While this article is primarily

concerned with a later period, remarks in the article furnish a basis for inferring the state of affairs prior to 1958; see, for example, pp. 9–10.

103. Quoted in ibid., pp. 8–9.

104. See generally Hughes and Luard, chaps. 5 and 6; and see Li Choh-ming, "First Decade."

105. Hughes and Luard, chap. 5.

106. Ibid., chap. 6.

107. *Hsin-hua pan-yueh-k'an* (New China semi-monthly), no. 24:57–59 (Dec. 25, 1957), tr. in Chao Kuo-chun, pp. 104–112; and see ibid., pp. 55–56.

108. Hughes and Luard, chap. 6.

109. "Take the Whole Country as a Coordinated Chess Game," *JMJP* (Feb. 24, 1959), tr. in *SCMP*, no. 1970:1–7.

110. Hughes and Luard, chaps. 12 and 13.

111. Li Choh-ming, "First Decade"; Wu Yuan-li, "Communist China's Industrialization at the Crossroads," *Current History* (September 1961), pp. 151–156.

112. Li Choh-ming, *Statistical System,* pp. 117–118 and 140–149.

113. Chao Kuo-chun, pp. 2–3.

114. "The Method of Drawing Up Annual Plans Undergoes An Important Reform," *Chi-hua ching-chi* (Planned economy), no. 9 (Sept. 9, 1958), tr. in *ECMM*, no. 152: 33–38; see also Yang Ying-chieh, "On Unified Planning and Phased Controls of National Economy," *Chi-hua ching-chi* (Planned economy), no. 11 (Nov. 9, 1958), tr. in *ECMM*, no. 156:24–28.

115. "Take the Whole Country," p. 2.

116. Ibid., p. 3.

117. Li Choh-ming, *Statistical System,* pp. 71 ff.

118. Ibid., p. 85.

119. Ibid., p. 145.

120. Ibid., pp. 111–149.

121. Quoted in Yang Ch'un-hsu and Li Cheng-jui, p. 8.

122.   Ibid., pp. 8 ff.; see also article from *TKP* (Jan. 18, 1959), tr. in *SCMP*, no. 2009:14–15.

123.   Yang Ch'un-hsu and Li Cheng-jui, p. 10.

124.   Ibid., p. 13.

125.   Ibid.

126.   Ibid., esp. pp. 17–18 and 20.

127.   Ibid., esp. p. 23; see also "Production Contracts Must Follow the Mass Line," *NFJP* (May 14, 1959), tr. in *SCMP*, no. 2025:17–20.

128.   Lin I-chou, "Contract of Production Must Be Practical," *Hung-ch'i* (Red flag), no. 10 (May 16, 1959), tr. in *ECMM*, no. 172:7.

129.   Yang Ch'un-hsu and Li Cheng-jui, "Apply the Contract System."

130.   Li Choh-ming, "First Decade."

131.   These provisions are discussed in "Carry out More Properly the Investment-by-Contract System," *JMJP* (Mar. 12, 1960), tr. in *SCMP*, no. 2220:3.

132.   Ibid.; see also Chi Tsung, "Preliminary Experience in the Contract System of Investment," *Ts'ai-cheng* (Finance), no. 9 (Sept. 5, 1958), tr. in *ECMM*, no. 147:68–71; and "Ministries of Metallurgical Industry and Finance Point Out at Conference that Contract Investment System Is the Correct Direction for Capital Construction Management," NCNA (Peking, July 14, 1958), tr. in *SCMP*, no. 1822:4–5; and see generally "Investment in Capital Construction Must Be Placed on a Contract System," *JMJP* (Peking, July 15, 1958), tr. in *SCMP*, no. 1822:2–4.

133.   "Carry out More Properly," p. 3.

134.   "Concluding Speech Made by Chin Ming, Vice-Minister of Finance at National Conference for the Exchange of Experiences Regarding the System of Investment-by-Contract for Capital Construction," *Ts'ai-cheng*, no. 5 (March 1960), tr. JPRS, no. 2984, Microfilm no.: Oct. 1960, 15585, pp. 1–21.

135.   See above.

136.   Chi Tsung, p. 70.

137.   "Concluding Speech Made by Chin Ming," p. 4.

138.   Hsia Yun, "Investment Contract System, A Revolution in the Management of Capital Construction," *Hung-ch'i*, no. 6 (Mar. 16, 1960), tr. in *ECMM*, no. 207:22–28; and see "Promote Investment-by-Contract System for Capital Construction in Shanghai," *Chieh-fang jih-pao* (Liberation daily) (May 5, 1960), tr. in JPRS, no. 5485:35–37.

139.   Chi Tsung, p. 69.

140.   "Rapidly Promote Investment-by-Contract System for Capital Construction," *Chieh-fang jih-pao* (May 5, 1960), tr. JPRS, no. 5495:38–39.

141.   "Speech by Wang Hsueh-ming, Vice-Minister of Finance, before the National Con-ference for Exchange of Experiences Regarding the System of Investment by Con-tract for Capital Construction on January 11, 1960," *Ts'ai-cheng,* No. 5 (Mar. 9, 1960), tr. in JPRS, no. 2984, Microfilm no.: 1960, 15585, pp. 24–36, esp. p. 31.

142.   "Ministries of Metallurgical Industry," p. 5.

143.   "Speech by Wang Hsueh-ming," p. 30.

III.   *Contract Law in China: A Comparative Analysis*

1.   See below.

2.   While there are compilations of decrees, etc., there does not appear to be a compre-hensive compilation of all relevant decrees on any topic. Moreover, it is not unlikely that many important laws are not published.

3.   See Chapter II, n. 24.

4.   Aside from the usual danger inherent in relying on any single source, there are addi-tional dangers at this time in relying on a source produced by the Chinese govern-ment and relating to the civil law. First, it may be merely propaganda. Second, even if it is not merely propaganda, lack of trained personnel to apply policies stated in the source may make it a nullity in effect. Third, in any event, the regime is still pressing the revolution and may not be willing to subject itself to law, and it may be said that on no account is it willing to adhere for long to any unchanging set of laws. Fourth, the law set forth in the source appears to be in good measure a carbon copy of Russian law, not something the regime has developed itself and, therefore, not something to which it has a strong commitment, especially since the Great Leap and the Sino-Soviet split. The facts suggest that while these dangers do exist they are not sufficient to preclude substantial reliance on this source for purposes of reaching ten-tative conclusions.

It appears unlikely that *Basic Problems* was compiled and published merely for the ends of propaganda. *Basic Problems* purports to have been compiled from a series of lectures given in early 1957 to students—prospective political-legal cadres—at the In-stitute of Civil Law, Central Political-Legal Cadres' School. It was revised and pub-lished in 1958 purportedly as "a reference for incumbent political-judicial cadres," *Basic Problems*, p. 7.

Thus, *Basic Problems* appears to have been written as a practical handbook to guide the enforcement of state policies throughout the country. In view of the stra-tegic importance assigned to the training of cadres and their role in implementing the policies of the regime, it is highly improbable, in view of the professed purpose of

Basic Problems, that it is mere propaganda. And, since the need for better-trained cadres has been acknowledged repeatedly by the regime, the publication of this book seems in fact to have been designed to answer that need.

The second danger—that the book will have no practical significance because of the lack of trained personnel to execute its policies—may have been a real one. However, this lack of trained personnel in China was not confined to the administration of law, and one might as well in earlier years have refrained from speaking of the role of statistical analysis for the same reason. Although the lack of such personnel impeded execution initially, it is likely that the situation improved, as it has in the realm of statistics. Moreover, the fact that the regime seems to support these policies has great importance without regard to their immediate effect in practice.

The third danger can be virtually discounted in light of the Russian and early Chinese experience. While it is clear that in the near future law will not assume the importance it has in the West, it is also clear that it has and increasingly will continue to perform necessary functions for the regime. As suggested above, the question of ultimate subjection of the regime to the rule of law will not arise in the ordinary cases, and in this sphere the role of law may be significant. The Chinese Communists' experiences have illustrated that law can be used to implement some of the goals of the revolution, particularly to realize the goals of economic planning. Although the specifics of "contract" law may be modified as the context changes, it is probable that set in the context of economic planning, the broad policies will remain relatively constant. Unless one believes that the "mass movement" element in the post-1949 Chinese development pattern provides an alternative mode of development, it seems reasonable to conclude that after the Cultural Revolution China has again returned to more routinized, bureaucratic means of development, which implies greater reliance on rules and rule-following.

Finally, the fact that the law presented in *Basic Problems* is in large part a transplanting of Russian law may cut both ways. Despite increased Chinese hostility to things Russian and China's rejection of particular elements of the Russian model for economic development, planning is still the keynote. To the extent China is relying less upon the highly centralized, planned physical allocations of resources than Russia did, China may need a more sophisticated legal system to facilitate decentralized methods of distribution.

Moreover, the Chinese Communists, it should be emphasized, adopted Russian law as it developed under economic planning, not simply as it appears in the 1922 code. And the Chinese have not accepted Russian law in every detail. The existence of a degree of selectivity indicates that the adoption of Russian law may be a conscious commitment. Of course, the Chinese regime will continue to mold its own actions to fit conditions in China, but unless it rejects the fundamentals of a planning model it seems unlikely to reject economic laws.

This commonsensical inference is buttressed by the ideological commitment to the proposition that law is a reflection of economic conditions, that economic conditions require overall economic planning and, therefore, that law should reflect the needs of economic planning.

5. See Vladimir Gsovski, *Soviet Civil Law* (Ann Arbor, 1948), I, 15–235, for a translation of the 1922 Code. It should be noted that technically it is a misnomer to call this code the Russian or Soviet Civil Code. As Gsovski points out in this volume,

pp. 4–6, this code is the code of one of the Republics, the R.S.F.S.R., of the Soviet Union. And, while many of the other republics have copied this code verbatim or with only minor changes, there appears to be no all-union code. In this latter connection see also Berman, "Soviet Law Reform," pp. 1199–1202.

6. For excerpts from the German Civil Code (BGB) of 1900 see Arthur T. von Mehren, *The Civil Law System* (Englewood Cliffs, 1957), appendix B, pp. 874–899. It should be noted that the laws of the Republic of China were also strongly influenced by German laws and therefore are quite similar in this respect. The possible relationship between Republican and Communist law is, however, not explored in this book.

7. Gsovski, I, 415.

8. *Basic Problems,* pp. 37–116.

9. Ibid., pp. 117–175.

10. Ibid., pp. 176–199.

11. Ibid., pp. 215–324.

12. Ibid., pp. 45, 69–72, 176 ff.

13. Ibid., p. 177; see ibid., pp. 65–77 for a discussion of the subjects of juristic relations.

14. Ibid., p. 69; see ibid., pp. 69–72 for a discussion of a citizen's disposing capacity.

15. See ibid., pp. 180–182 for a discussion of these three sources of obligations.

16. See excerpts quoted in von Mehren, *Civil Law System,* pp. 463–464.

17. Ibid.

18. Gsovski, I, 424.

19. With regard to both German and French law, see Arthur T. von Mehren, "Civil Law Analogues to Consideration: An Exercise in Comparative Analysis," *Harvard Law Review* 72:1009 (April 1959).

20. In light of the discussion in ibid., it appears, for example, that the evidentiary and cautionary functions of consideration as a formality in effect are performed by direct formal requirements. In this regard see Gsovski, I, 431–432. With respect to the function of screening individual transactions for unfairness, see Gsovski, I, 425, with reference to the right of avoidance in situations where a party "under the pressure of distress" has entered a contract "clearly unprofitable to him."

21. The function of consideration at common law to screen individual transactions for

gross unfairness is performed by a direct right to redress in Chinese law. See above, p. 44, for reference to the right of an oppressed party to avoid a contract that is obviously unfair if that party was in an unfair bargaining position when he concluded the contract.

22.   *Basic Problems*, p. 204.

23.   Ibid., pp. 204–208; and for a brief discussion of the comparative practice of "freedom of contract" in Western and Soviet law see Harold J. Berman, *The Nature and Functions of Law* (Brooklyn, 1958), pt. 3, esp. pp. 93–96.

24.   See below.

25.   *Basic Problems*, p. 203.

26.   Ibid., pp. 203–204.

27.   BGB, Articles 145–148, tr. in von Mehren, *Civil Law System*, p. 881.

28.   Ibid.; see also excerpts in ibid., pp. 492–493 regarding the effect of the clause "freibleibend."

29.   Ibid., pp. 492–493, discussing the impact of the rules regarding offer and acceptance in the common law, French law, and German law.

30.   *The Russian Civil Code*, section 132–135, tr. in Gsovski, II, 111–112.

31.   *Basic Problems*, pp. 180, 236.

32.   Ibid., p. 213.

33.   Yang Ch'un-hsu and Li Cheng-jui, "Apply the Contract System."

34.   See discussion in von Mehren, *Civil Law System*, pp. 472–473.

35.   *Restatement of the Law of Contracts* (St. Paul, 1933), sec. 12 and comments thereto.

36.   *Basic Problems*, pp. 258, 282–283.

37.   *Restatement*, sec. 45.

38.   *Basic Problems*, pp. 258, 282–283.

39.   Ibid., p. 281.

40.   For example, see below for reference to limitation of carrier's liability for breach of contract.

41.   von Mehren, *Civil Law System*, p. 473.

42.   *Basic Problems*, p. 211. It is interesting to note that the reasoning in the section of *Basic Problems* dealing with terms of contracts, pp. 208–211, is not unlike that in the famous common law cases of *Scammel v. Ouston*, in the *House of Lords* [1941] 1 All E.R. 14, and *Foley v. Classique Coaches, Ltd.* [1934] 2 K.B. 1, appearing in Lon L. Fuller, *Basic Contract Law* (St. Paul, 1947), pp. 76 and 81 respectively.

43.   *Basic Problems*, pp. 200–213.

44.   Ibid.; for a brief discussion of "the basic conditions of supply" in Russia, see Berman, *Justice*, p. 71; and see Berman, "Commercial Contracts in Soviet Law," *California Law Review* 35:191–225 (June 1947), at pp. 201–202.

45.   *Basic Problems*, pp. 227–232.

46.   Ibid., p. 234.

47.   Ibid., p. 201; see below for a discussion of penalties.

48.   Ibid., p. 80.

49.   *The Russian Civil Code*, section 137 (a), tr. in Gsovski, II, 112.

50.   *The Russian Civil Code*, section 135, ibid., p. 112.

51.   *Basic Problems*, p. 136.

52.   Ibid., pp. 79, 213–214.

53.   Ibid., pp. 278 and 80.

54.   *The Marriage Law of the People's Republic of China, Article 6*, tr. in C. K. Yang, *The Chinese Family in the Communist Revolution* (Cambridge, Mass., 1959), p. 222.

55.   *Basic Problems*, pp. 80–81.

56.   Berman, "Commercial Contracts," p. 210. The parol evidence rule excludes evidence derived from negotiations prior to the reduction of an agreement to writing where such evidence contradicts the written agreement if the latter appears to encompass the entire bargain between the parties.

57.   Gsovski, I, 425.

58.   For a brief discussion of uniqueness and specific performance at common law, see Fuller, *Basic Contract Law*, pp. 27–31.

59.   See Berman, *Nature and Functions*, pp. 390 ff.

60.   *Restatement,* sec. 358 ff.

61.   Ibid., sec. 347.

62.   Ibid., sec. 275.

63.   See below.

64.   See discussion in von Mehren, *Civil Law System,* pp. 501–505.

65.   For example, by the mid-1930s Mao had evolved a strategy whose basic features, at least by implication, were homegrown adaptations of Marxism-Leninism. For a fine and detailed discussion of this see Benjamin I. Schwartz, *Chinese Communism and the Rise of Mao* (Cambridge, Mass., 1958), esp. pp. 189–204.

66.   *Basic Problems,* pp. 180–181.

67.   Ibid., pp. 180, 236.

68.   For example, ibid., p. 281.

69.   See above.

70.   *Basic Problems.*

71.   Ibid.; for a brief discussion of general, local, and direct contracts in Soviet law see Vladimir Gsovski and Kazimierz Grzybowski, *Government, Law and Courts in the Soviet Union and Eastern Europe* (New York, 1959), II, 1149–50.

72.   *Basic Problems,* p. 236. It is important to note that it is not clear which administrative organs decide such disputes. So long as the disputes are between enterprises that are under the same specialized industrial company or the same ministry, the problem of which organ decides is relatively unimportant. However, since there is no public arbitration organ in China comparable to Gosarbitrazh in Russia (see *Basic Problems,* p. 214; and Berman, *Justice,* pp. 63 ff.), it is uncertain how disputes between enterprises under different ministries will be settled where the ministries themselves are unable to agree. It seems unlikely that the State Council, the only executive body above the ministries, would take the trouble to settle all such disputes itself.

73.   *Basic Problems,* p. 228.

74.   Ibid., p. 241.

75.   Ibid., pp. 242–244 and 191.

76.   Ibid., pp. 231–232.

77.   Z. L. Zile, "Remedies for Breach of Contractual Obligations in Soviet Law" (student paper, The Harvard Law School, Comparative Legal Research, 1960), p. 14.

78.   *Basic Problems*, pp. 321–332.

79.   Ibid., p. 240.

80.   Ibid., p. 185.

81.   Ibid., pp. 240–241.

82.   Ibid.

83.   Zile, pp. 12–13.

84.   *Basic Problems*, p. 193.

85.   Ibid., pp. 192–193.

86.   Ibid.

87.   Ibid., p. 241.

88.   Ibid., p. 193.

89.   Ibid., pp. 185–187; see below for discussion of "contractual fault," "*force majeure*," and "impossibility."

90.   See discussion in *Basic Problems*, p. 202.

91.   Mihaly, pp. 323–324.

92.   *Basic Problems*, pp. 193–194.

93.   Ibid.

94.   Zile, p. 38.

95.   *Basic Problems*, p. 193.

96.   This notion is suggested with regard to the parallel Russian law in Zile, pp. 38–39.

97.   *Basic Problems*, p. 193.

98.   *Restatement*, sec. 339.

99.   *Basic Problems*, pp. 186–189.

100. Ibid., p. 241.

101. Ibid., p. 185.

102. Ibid., p. 326.

103. With regard to the emphasis on "law as a teacher and parent" in Russia, see Berman, *Justice,* pp. 203–206 ff.

104. Under the expectancy measure, "The law attempts to put the plaintiff in as good a position as he would have occupied had the defendant performed his obligation." von Mehren, *Civil Law System,* p. 507; and see *Restatement,* sec. 329 ff.

105. *Basic Problems,* p. 187.

106. See generally Lon L. Fuller and W. R. Perdue, "The Reliance Interest in Contract Damages," *Yale Law Journal* 46:52 (November 1936).

107. *Basic Problems,* p. 187.

108. Ibid., p. 188.

109. Ibid., pp. 188, 335.

110. Zile, p. 54.

111. See *Basic Problems,* pp. 188, 335.

112. *Basic Problems,* p. 290.

113. Ibid., pp. 91, 196, 340.

114. Ibid., pp. 185–187.

115. For the definition and rules regarding impossibility at common law see *Restatement,* sec. 454 ff.

116. Such a conclusion does not seem unreasonable under the terms of ibid. sec. 460, particularly illustration no. 12 in the comments thereto.

117. Berman, "Force Majeure and the Denial of an Export License under Soviet Law: A Comment on 'Jordan Investments, Ltd. v. Soiuznefteksport," *Harvard Law Review* 72:1128 (April 1960).

118. *Basic Problems,* p. 238.

119. Ibid., pp. 84–87.

120.   BGB, article 138, tr. in von Mehren, *Civil Law System*, p. 181.

121.   *Basic Problems*, pp. 87–88.

122.   Ibid.

123.   Ibid., pp. 83–84.

124.   Ibid., p. 83.

125.   See Gsovski and Grzybowski, pp. 1140–44.

126.   *Basic Problems*, pp. 88–91.

127.   Ibid.

128.   See discussion in Gsovski, I, 426–431; and see Berman, "Commercial Contracts," pp. 211–212.

129.   Gsovski, I, 426.

130.   See above.

131.   See n. 128.

IV. *Contracts in Chinese Agriculture: A Case Study of Contracts in Practice, 1949–1963*

1.   The evidence often is scattered, and also may carry a bias tending to stress the state's role in agricultural contracts. No evidence has been found, for example, of contracts in which the state has not played some role. This has led me to conclude that such contracts either do not exist or are insignificant.

2.   To some degree, contracts everywhere involve similar contradictions; see Chap.V. I was provoked to pursue the issue of contradictions by Franz Schurmann's brilliant book *Ideology and Organization in Communist China* (Berkeley, 1966). In this chapter, however, I do not use the concept of contradiction in the highly technical way Schurmann does; see, for example, pp. 75–76 and 101–103 in his book. Schurmann's usage is discussed in my review-article of his book in *Pacific Affairs* 39:349–360 (fall and winter 1966–1967).

3.   See, for example, Jen Chieh-hsin, "Strengthen the Economic Contract Work and Promote the Smooth Execution of National Economic Planning," *Cheng-fa yen-chiu* (Political-legal research), no. 1:31 (1957); "How to Develop Contract Management," *JMJP* (Apr. 1, 1950); Ts'ui Li-ch'un, "Several Problems That Should Be Attended to and the Function of the Contract System with Regard to Rural Sideline Production

and Marketing," *Chung-kuo nung-pao* (Chinese agricultural news) 1.6:67 (1950); Li Ch'eng-shui et al., "Broadly Carry Out the Contract System," *TKP* (Feb. 20, 1959).

4. The 1954 Charter of the All-China Headquarters Supply and Marketing Cooperative in Peking reflects the degree to which these cooperatives are organs for carrying out state policy. In that charter the major tasks of the cooperatives are listed as the planned organization of buying and selling in the countryside, the expansion of exchange between city and village, the development of cooperative commerce and, implicitly, the displacement of private commerce, the promotion of the cooperativization of agriculture, and the raising of the standard of living of the SMC members. *JMJP* (Nov. 30, 1954). The SMCs form a hierarchy in which the lower levels are subject to control from above and in which SMCs at all levels are subject also to control by relevant state and party organizations of the same or higher levels.

5. In 1953, for example, about 75 per cent of total purchases by the SMCs was for the state. Ch'eng Tzu-hua, "Struggle to Further Develop the Exchange of Materials between City and Town, to Promote the Movement to Increase Agricultural Production, Taking Mutual Aid and Cooperation as its Core, and to Aid Industrialization of the Country," *JMJP* (Nov. 30, 1954).

6. Liu Ch'i-shan, "The Combined Contract is a New, Independent Contract System," *Cheng-fa yen-chiu,* no. 3:37–40 (1957). It should be noted, however, that there are a number of variant forms that have appeared at different times; e.g., *t'ung-kou, p'ai-kou* and *i-kou* contracts. It is very difficult, and perhaps not worth the trouble to distinguish among all the forms. As Gene Hsiao points out in his article "The Role of Economic Contracts in Communist China," *California Law Review* 53.4: 1052 (October 1965), the *t'ung kou* and *p'ai-kou* contracts generally involve goods in categories one and two respectively, while the *i-kou* contract involves goods in category three. What is said about the advance purchase contract appears to be relevant to *t'ung-kou* and *p'ai-kou,* and what is said of the combined contract seems relevant to *i-kou.*

7. Li Che-jen, "The Combined Contracts of the SMCs, APCs, and Mutual Aid Teams of Two Villages," *Hsin-hua yüeh-k'an* (New China monthly; May 1952), pp. 107–108.

8. Liu Ch'i-shan.

9. "The Directive Issued by the Cabinet for This Year's Advance Purchase of Cotton," *JMJP* (Mar. 23, 1955).

10. "The Directive Issued by the Cabinet for the Work of the Advance Purchase of Cotton," *JMJP* (Dec. 23, 1955).

11. "Unified sales" is a system of rationed distribution applied to important materials.

12. Liu Ch'i-shan, p. 40.

13. With regard to the advance purchase contract there is evidence that the monetary

advances have been made. This tends to support the position that a priority is involved. See "Directive of the Headquarters Branch of the People's Bank and the Grain Ministry Regarding Doing Well the Work of Supplying Capital for Summer Edible Oil Purchases," *Jen-min shou-ts'e* (People's handbook; Peking, 1962), p. 214. On the other hand, there is also evidence that the relationship between the parties may involve a fixed advance in kind to the agricultural producer, an advance that is arranged simultaneously with the signing of the contract, if not as a part of the contract itself. The purpose of this arrangement is to insure that the advance made will be employed in a manner that will benefit production. In this case it appears unlikely that the agricultural producer has an option to buy or not to buy. See "The Making of Advance Deposits in Advance Purchase of Cotton Must Be Combined with the Supply of Materials," *TKP* (Peking, Apr. 24, 1963). In any event, in light of the urgent needs of the peasants and the limited range of goods involved in advances in kind, in practice it is very likely that priority rights will always be exercised and that the end result in both the advanced purchase and combined contracts may be nearly the same.

14. Liu Ch'i-shan, p. 40.

15. "Several Problems in the Contract System after Communization," *Cheng-fa yen-chiu*, no. 4:34 (1959); Li Ch'eng-shui et al.; Liu Yung-hua (pen-name, Lou Hua), "The Lock and Chain on the Neck of the Mainland Small Trader—Discussing the Chinese Communists' Cooperative Undertaking of the Supply and Marketing 'Combined Contract' System," *Tsu-kuo chou-k'an* (Motherland weekly) 12.2:16 (Oct. 10, 1955). Since this magazine is published in Hong Kong by an organization with a known anti-Communist bias, it is perhaps proper at this point to suggest that, the pejorative character of the title notwithstanding, most articles written by Liu Yung-hua are reasonably objective, well-reasoned, and often stimulating.

16. See below for a discussion of retail trade in these commodities in the rural markets (*chi-shih*).

17. "Hsi Liang Large Brigade Signs an Advance Purchase Contract for Cotton with the State," *TKP* (Peking, Feb. 2, 1963).

18. "It Is Necessary to Seasonably Arrange the Work of Advance Purchase of Cotton," *JMJP* (Dec. 23, 1955).

19. Liu Yung-hua, "Lock and Chain," p. 13.

20. Liu Ch'i-shan, "The Combined Contract." The contract is said to "combine" (1) purchases of agricultural and sideline commodities with (2) sales to the agricultural production unit of material it needs for production and livelihood. The contract thereby indirectly combines the production and consumption plans of the agricultural production unit with the supply and marketing plans of the SMC. In light of the hierarchical structure of the SMCs and their subordination at all levels to state organs of commerce, agriculture, light industry, etc., ideally these contracts also

ultimately "combine" the basic agricultural production unit's plans with the needs of the relevant area's economy.

21.   "Extend the Combined Contract between the Mutual Aid Teams and the Cooperatives," *JMJP* editorial (Apr. 10, 1952). The first three clauses of the model contract
provide that the SMC is to give the mutual aid team priority in services and supply,
to favor the mutual aid team in marketing its native products that are difficult to
sell, and to arrange needed loans for the mutual aid team through credit units. Under the fourth and fifth clauses the mutual aid team promises to use the SMC exclusively for its sales and purchases. The team also promises that, in order to coordinate sales and purchases, at the middle of each month it will present its plans for
the following month's needs and sales as the basis for signing concrete executing
(*chü-t'i chih-hsing*) supply and marketing contracts with the SMC. The seventh
clause provides that the SMC will help the team make its plans and will participate
in meetings of the team. The commercial unit, thereby, is given a lever for directly
influencing the formation of the production unit's plans.

22.   Generally paralleling the geographic pattern of the Communist liberation, the combined contract began, and was most widely developed during the early years, in the
provinces of north and northeast China. It is said to have developed in Yung Chi
County of Kirin province a year or two before it was officially spread. Its real development, however, came after the April 10, 1952, *JMJP* editorial referred to above.
Based on incomplete statistics for north China, it appears that some 2,500 basic-
level SMCs concluded combined contracts with over 90,000 mutual aid teams and
APCs in 1952. But it is clear that these early contracts were on the whole unsuccess-
ful; in many areas it appears contracts were signed as a mere formality in order to
satisfy official exhortations. Often the cadres of the SMCs were not themselves
clear as to the nature of the combined contract. Performance was spotty and varied
greatly from place to place. For typical sources from 1952–1956 see: "The 1952
Plan and the Contracts Concluded between the Erh Tao K'ou SMC . . . and the
Hsieh Yun Hu Mutual Aid Team," *JMJP* (Apr. 10, 1952); "Actively and Steadily
Extending the Combined Contract is One of the Most Powerful Measures for Pro-
moting the Socialist Transformation of Agriculture," *JMJP* (June 28, 1954); "Con-
tinue to Implement the Supply and Marketing Combined Contract System," *NFJP*
(Nov. 9, 1954); "Our Cities Develop in Large Amounts the Combined Contract for
Vegetables," *Lü-ta jen-min jih-pao* (Lü-ta people's daily; Apr. 4, 1955 [?]); "Spread
the Contract System and Do the Work of Supplying Production Materials Well,"
TKP (Tientsin, May 13, 1956); "Widely Spread and Realistically Perform Combined
Contracts," *Yunnan Daily* (June 18, 1956); and "179 Production and Marketing
Vegetable Combined Contracts Were Concluded in the City: A Great Deed in the
Work of Reforming the Production, Supply, and Marketing of Vegetables," *Chang-
chiang jih-pao* (Changkiang daily; Nov. 21, 1956).

23.   *CNA,* no. 199:2 (Oct. 4, 1957).

24.   Liu Ch'i-shan, p. 38.

25.   Ibid.

26. "The Purchase Price Is Not Sufficiently Logical; Some APCs Do Not Perform Their Contracts," *TKP* (Peking, Jan. 4, 1957).

27. "The Cabinet Clearly Fixes the Scope of the Free Market," *TKP* (Peking, Aug. 18, 1957). I have not found any evidence supporting the popular notion that the free markets actually were closed at this time. Interviews indicate free markets continued to exist up to communization, when they were starved out of existence. This conclusion appears to be supported by the statement late in 1958 that "the free market basically no longer exists." "Decision of the Central Committee of the Chinese Communist Party and the Cabinet, Regarding Improving the System of Rural Finance and Trade Administration in Response to the Conditions of Communization," *Hsin-hua pan-yueh-k'an* (New China semi-monthly; Nov. 1, 1959), p. 63.

28. "Purchase (*p'ai-kou*) Work Can't Be Simplified," *NFJP* (Mar. 20, 1963).

29. "SMCs and APCs Should Conclude Combined Contracts," *Che-chiang nung-ts'un kung-tso t'ung-hsün* (Chekiang agricultural work bulletin), no. 49 (Mar. 4, 1955) as reprinted in *Chung-kuo nung-ts'un ti she-hui-chu-i kao-ch'ao* (Socialist high tide in Chinese agriculture; Peking, 1956), middle volume, pp. 679–680.

30. Li Ch'eng-shui et al., "Broadly Carry Out."

31. Ibid.

32. Wei Fu-k'ai et al., "Several Problems Regarding Implementing the Contract System," *TKP* (Peking, June 5, 1959).

33. *TKP* (Mar. 4, 1959); Li Ch'eng-shui, "Broadly Carry Out."

34. "Introduce the Connecting Contract between the An Yang Wu Kuan Supply and Marketing Cooperative, Credit Cooperative and Agricultural Producers' Cooperative," *TKP* (Tientsin, May 25, 1954); "The Supply and Marketing Cooperative of Lu Ch'ing County, Wei-tzu Town and Concerned Departments Closely Co-ordinated (*p'ei-ho*) to Carry Out the Four-way Combined Contract to Help Agricultural Production," *TKP* (Peking, Feb. 25, 1962); "Chin-hua Wholesale Station, Concluding Order Contracts with Local Handicraft Producers' Cooperatives, Improved the Production of the Producers' Cooperative and Fulfilled the Supply and Marketing Cooperatives' Requirements for a Source of Goods," *TKP* (Tientsin, Nov. 13, 1953).

35. On the criminal side, see Jerome Alan Cohen, "The Criminal Process in the People's Republic of China: An Introduction," *Harvard Law Review* 79.3:469–533 (January 1966). And with regard to the disposition of civil disputes, see generally Cohen, "Chinese Mediation on the Eve of Modernization," *California Law Review* 54:1201–26 (August 1966).

36. "Provisional Measures of the Finance and Economic Committee of the Central People's Government Affairs Council Regarding the Signing of Contracts Between Organs, State-Managed Enterprises, and Cooperatives," *Chung-hua jen-min kung-ho-kuo*

fa-ling hui-pien (1949–1950) (Collection of laws and decrees of the Chinese People's Republic; Peking, 1953), p. 532. It appears that these provisional regulations were still in effect at least as late as 1957. Jen Chien-hsin, p. 33. A similar pattern for settlement of disputes arising from processing and order contracts in industry was referred to in 1955 by Tseng Shan, then Minister of Commerce, "Special Problems Regarding Processing (*chia-kung*) and Orders (*ting-huo*)," *Kung-shang-chieh yueh-k'an* (Industrial and commercial monthly; Mar. 10, 1955), p. 8; and see Hsieh Wu, "Discuss the Contract System," *Cheng-fa yen-chiu,* no. 2:43 (1959).

37. Ibid., and "Provisional Measures of the Finance and Economic Committee of the Central People's Government Affairs' Council."

38. It is interesting to note that in the Soviet Union failure in the state supply apparently does not excuse a party from performance of his contract obligation. Dietrich A. Loeber, "Plan and Contract Performance in Soviet Law," University of Illinois *Law Forum* (spring 1964), pp. 148–50.

39. "Several Problems in the Contract System After Communization," p. 36.

40. "Agriculture and Commerce Conclude 'Five Fix' Production and Marketing Contracts," *JMJP* (May 3, 1961); Ts'ui Li-chun, "Several Problems."

41. "The Purchase Price is Not Sufficiently Logical"; "Pay Attention to Contract Management," *TKP* (Peking, 1957), in Union Research Institute files, 1957–1958, vol. 84, no. 32165.

42. It is suggested that the peasants must be "educated" every year about the solemn nature of contracts. "Purchase Work Can't Be Simplified."

43. "Spread the Contract System and Do the Work of Supplying Production Materials Well"; Yen Wang-fu, "How to Realize (*tui-hsien*) the Purchase and Marketing Combined Contract," *TKP* (Peking, Mar. 30, 1962); "T'ai Yuan City Supply and Marketing Cooperative, After Investigation and Finding out the Causes for Problems that Existed in Executing Contracts, Fixed Improved Methods," *TKP* (Tientsin, June 10, 1953).

44. "Settle (*ch'ing-li*) Combined Production and Marketing Contracts for Live Pigs," *Che-chiang jih-pao* (Chekiang daily; July 15, 1955).

45. Jen Chien-hsin, pp. 32–33; Liu Ch'i-shan, p. 39. Many contracts, particularly in the early years of the First Five-Year Plan, were not performed seriously. Since then, performance appears to have improved, especially as to first-category goods, whose production and sale is carefully regulated. Statistics on performance generally are scattered and probably unreliable. They are suspiciously consistent in indicating for any given time that, though performance in the past has not been very good, recently performance has improved greatly. See for example, Yen Wang-fu, "How to Realize"; and "In Purchasing Agricultural and Sideline Products," *TKP* (Feb. 15, 1962).

46.   Dwight H. Perkins, *Marketing Control and Planning in Communist China* (Cambridge, Mass., 1966), pp. 28–42, 199.

47.   With the commune movement in the last half of 1958 and the transfer of the basic-level government commercial organs and SMCs to commune administration, the SMCs whose scope already had been restricted by 1958, entered a period of enforced hibernation lasting until the first half of 1961. *CNA* (Aug. 31, 1962), pp. 2–3. For a short period, until the end of 1958, it was thought that the communes should be self-sufficient and commerce reduced accordingly. See "Decision . . . Regarding Improving the System of Rural Finance and Trade Administration." Since parties to contracts concluded immediately after communization were much larger units, commonly the commune and the country commercial department, the scope of the contract seems to have expanded in 1959. Whereas in 1957 there is evidence that it was believed the length of the contract period should depend in part on the nature of the product involved, just after communization the emphasis generally appears to have shifted to year-long contracts. "The Ch'iao (?)–k'ou SMCs Spreading the Combined Contract Is Not a Coincidental Matter," *Hsin Hu-nan pao* (New Hunan daily; May 15, 1957); "The Lung Chang Commercial Bureau and the Hu Chia Commune Conclude Purchase and Marketing Contracts; the State Plan Is Bound to the Commune Plan," *TKP* (Peking, Nov. 2, 1958). Similarly the contracts of late 1958 and 1959 incorporated the broad needs and produce of the diversified economy of entire communes. This change in the scope of contracts reflects the fact that with the abolition of private plots whatever was not included in plans or contracts might not be produced or sold. Since, practically speaking, the scope of centralized planning at best was limited to important products, combined contracts had to provide direction for products outside this scope. Li Ch'eng-shui, "Broadly Carry Out."

48.   Editorial and articles in *TKP* (Apr. 4, 1959).

49.   Liu Yung-hua (pen name, Yu Heng), "Commenting on the Chinese Communists' Rural Market Trade," *Tsu-kuo chou-k'an* (Hong Kong, Nov. 2, 1959), p. 21.

50.   "The Cabinet Approves and Transmits the Report of the Ministry of Commerce Demanding Level-by-level Convoking of Materials-Exchange Conferences," *Hsin-hua pan-yueh-k'an*, no. 7 (1959).

51.   See, for example, Ts'ui Chih-min, "Expand the Purchase and Marketing of Native Products, and Extend Contract Management," *Chin-pu jih-pao* (Advance daily; Tientsin, Feb. 24, 1951); and "The Advantages and Understanding of Convoking Combined Conferences for Supply, Production, and Marketing," *Hsin-chiang jih-pao* (Sinkiang daily; Aug. 27, 1955).

52.   "The Cabinet Approves and Transmits . . ."

53.   These conferences are to be held twice a year, in the spring to make the contracts and organize production, and in the autumn to examine contract performance and, if necessary, to supplement and revise the contracts to make them more realistic. The

conferences may deal with a specific type of product, like the National Native Waste Products Conference at which over 900 contracts and agreements valued at over U.S. $120,000,000 were signed. "Holding Materials-Exchange Conferences Strengthens the Planned Nature of Production and Circulation," *TKP* editorial (Mar. 23, 1959). Or they may deal with an even wider range of products like the first National Conference for the Exchange of Third-Category Materials, at which over 22,000 contracts and agreements valued at almost U.S. $400,000,000 were signed. "The First Whole Country Materials-Exchange Conference for Third Category Materials," *JMJP* (Mar. 23, 1959), as reprinted in *Hsin-hua pan-yueh-k'an*, no. 11:127 (1959); "Bringing the Production and Marketing Activity of the Commune Production Brigade into State Plans, Various Places Widely Sign Production and Marketing Contracts," *JMJP* (Apr. 13, 1959). See also Hsiao, pp. 1055–56.

54. This emphasis is in line with the slogan "take the whole country as a chessboard," a slogan spread in early 1959 as part of a campaign to combat localism that resulted from the extreme decentralization of 1958.

55. *CNA*, no. 299:5 (Oct. 30, 1959). This issue is very useful generally regarding contracts, commerce, and institutional changes in 1959.

56. "The Directive of the Central Committee of the Chinese Communist Party and the Cabinet Relating to Organizing Rural Market Trade," *TKP* (Peking, Sept. 25, 1959).

57. Kuan Ta-t'ung, "My Country's Unified Socialist Domestic Market," *Hung-ch'i*, no. 6: 28–35 (1963), esp. after p. 30.

58. Kuan Ta-t'ung, "Discuss Rural Market Trade," *JMJP* (Nov. 21, 1959). For a highly refined view of rural markets see G. William Skinner, "Marketing and Social Structure in Rural China," *Journal of Asian Studies* 24.1:3–43 (November 1964); 24.2: 195–228 (February 1965); 24.3:363–399 (May 1965). See below for my discussion of the relation between Skinner's "standard market" and what I refer to as rural markets. At least in "standard markets," and for the most part in rural markets in general, contracts do not seem to play a role in the market proper.

59. In the case of first- and second-category goods, the state apparently can pre-empt their sale in the rural market if necessary even though previously set state tasks have all been fulfilled. "The Directive of the Central Committee of the Chinese Communist Party and the Cabinet Relating to Organizing Rural Market Trade."

60. A number of reasons have been postulated to explain the delay. Some relate to conditions at the center; others to local conditions; still others involve conditions directly relevant to both. First, it seems likely there was a significant division of opinion among the central leadership regarding what should be done. This is reflected in the fact that most of the measures before 1961 to reorganize the economy were only "half measures." Perkins, p. 93. It is also suggested by the apparent division of opinion within the top leadership in 1958 on the original question of communization. Schurmann, *Ideology and Organization*, pp. 474–477, 483–484, 490–492.

It seems unlikely that the major policy shifts between 1956 and 1958 could have occurred without the prior or subsequent formation of what Schurmann refers to as "opinion groups;" pp. 55–56. Second, the extreme decentralization of controls, such as statistical controls, meant that central, and even provincial- and lower-level leadership lacked reliable information about conditions in many areas. Perkins, p. 90. Third, the decentralization of all authority to rural Party cadres, previously radicalized by recent rectification campaigns and utopian propaganda and localized by campaigns stressing self-sufficiency, undoubtedly meant that sustained efforts would have had to have been made by central authorities to carry out a reorganization. The rural cadres, as Skinner, pp. 373–374, points out, had been convinced that commerce was inherently evil and bourgeois. Given this orientation, economic reorganization of the sort required demanded either a reassertion of central controls or a fundamental change in the viewpoint and level of expertise of rural cadres. Neither alternative could be implemented very quickly. Finally, if one takes the marketing institutions as form, and the products with which they deal as content, form could not take on content until the harvests, which were the source of most of the products marketed, improved significantly.

61. Kuan Ta-t'ung, "Discuss Rural Market Trade"; "To Guide the Communes to Carry out Planned Production," *TKP* (Jan. 19, 1959); Liu Yung-hua, "Commenting on the Chinese Communists' Rural Market Trade," p. 20; "Bringing the Production and Marketing Activity of the Commune Production Brigade into State Plans." Scattered evidence indicates that in 1959 and 1960 in some areas contracts were being signed directly with production brigades instead of with the larger commune unit. It appears that by the end of 1960 the rule was that the contracts were to be signed with the production unit directly concerned, either the commune, the (large) brigade, or the team (small brigade). "Several Problems in the Contract System After Communization"; and "The Commercial Departments in Wan-hsien Special District Completely Carry Out the Contract System," *TKP* (Peking, Mar. 22, 1961). This shift may have predated somewhat the official reorganization of the commune in the winter of 1960–1961, under which successive retrenchment to the brigade and, later, to the team was carried out.

62. Anna Louise Strong's exceptionally interesting article in the June 17, 1964, issue of the *Peking Review* denies that the commune unit has been abandoned in practice.

63. Liu Yung-hua, "Communist Chinese Domestic Trade in 1961," *Tsu-kuo chou-k'an* (Hong Kong, Apr. 30, 1962), pp. 12–17. At the various materials-exchange conferences tens of thousands of contracts and agreements, valued at hundreds of millions of dollars, were signed. Both within and outside the materials-exchange conference system the basic-level SMCs were back in their role of directing production and sales through contracts. "In Purchasing Agricultural and Sideline Products"; "The Purchase and Marketing Combined Contract is Good," *TKP* (Jan. 30, 1962).

64. In some areas, at least, the conferences were held level-by-level as originally envisaged, thus facilitating organized exchange.

65.   Liu Yung-hua (Lou Hua), "Chinese Communist Domestic Trade in 1963," *Tsu-kuo chou-k'an*, no. 2:14–18 (1964); "Widely Connecting, Actively Selling, Signing Contracts to Guide Production," *TKP* (Jan. 11, 1964).

66.   Skinner, n. 58 above.

67.   Ibid., p. 379.

68.   Although I am not entirely clear about Skinner's terminology, I take it that the rural market (*chi-shih*) is equivalent to Skinner's generic "traditional periodic markets" (p. 379). Of the 42–45,000 such traditional periodic markets Skinner estimates as existing in China today (pp. 376, 379), 32–34,000 are said to be standard markets in effect. The remainder, I presume, are unmodernized intermediate and central markets.

69.   Ibid., p. 6.

70.   "The funneling and supply functions of the intermediate market town and the bulking and wholesaling functions of the central market town are [today] . . . shared by a variety of agencies. . . The local production of 'first-category' items . . . eventually reach the procurement stations which state companies maintain in the intermediate market towns, and are shipped on to their warehouses in central market towns and cities. Other local products of the 'second category' . . . are, in the areas of their major production, likewise handled solely by 'state commerce.' First- and second-category items for local consumption are imported into a given trading system solely by the state trading companies, which distribute them through warehouses maintained in central market towns, and through wholesaling agencies of the supply and marketing co-operatives situated in both central and intermediate market towns. Retail outlets for controlled commodities of this kind . . . are either the [supply and marketing] co-operatives' own stores in the market towns or petty traders selling on their behalf. Finally, in the case of . . . 'third category' . . . goods . . . vertical distribution is effected in part through a special type of warehouse under state commerce which 'acts as intermediate link and accepts responsibility for storing, buying and selling of such commodities. . . Their purpose is to channel goods from the villages to the cities.' Vertical distribution of third-category items is also accomplished through the flourishing institution now known as the [materials-exchange conference]." Ibid., p. 380.

71.   Ibid., pp. 380–381.

72.   Liu Yung-hua, "Chinese Communist Domestic Trade."

73.   John B. Quigley, Jr., "Recent Comment on the Contract of Delivery by Soviet Writers," spring 1966, unpublished paper at the Harvard Law School; and Loeber, "Plan and Contract."

74.   Quigley, pp. 2–3; but see Loeber, p. 173.

75. Quigley, p. 7. The *nariad* is one kind of planning act, specifically a delivery order issued by an agency administratively superior to the recipient enterprise.

76. Loeber, pp. 146–148, 155–159. Arbitrazh is a quasi-judicial, hierarchical network within government with exclusive jurisdiction for administratively deciding disputes between socialist economic organizations.

77. Quigley, p. 18.

78. Ibid., pp. 2, 14–15, 20.

79. Schurmann, *Ideology and Organization*, p. 178.

80. Ibid., pp. 175–176.

81. Ibid., pp. 219–297; and see Franz Schurmann, "China's 'New Economic Policy'— Transition or Beginning," *The China Quarterly*, no. 17:65–91 (January-March 1964).

82. Schurmann, *Ideology and Organization*, p. 219.

83. Schurmann, "China's New Economic Policy," pp. 76–80.

84. Schurmann postulates that China may have been going through an NEP from 1961 to at least early 1963. Ibid., pp. 80–83.

85. Ibid., p. 77.

86. The *tolkach* or "pusher" is a "supply expediter . . . sent out by a firm to 'push' for its interests." Joseph S. Berliner, "The Informal Organization of the Soviet Firm," *The Quarterly Journal of Economics* 66:358 (August 1952).

87. If the materials-exchange conference system and analogous institutions in the industrial realm are primarily facilities for implementing decentralization I, such face-to-face confrontation by production, distribution, and user units would reduce greatly the need for advertising as a precondition for widespread direct contracting.

88. Schurmann, *Ideology and Organization*, p. 73.

89. For another, but not dissimilar, approach to the more general problem of "commandism" and consensus, see John Wilson Lewis, *Leadership in Communist China* (Ithaca, 1963).

V. *Contracts in America in Perspective*

1. Short critiques of Maine's theory can be found in Max Radin, "Contract Obligation and the Human Will," *Columbia Law Review* 43:575–585 (July 1943); Nathan

Isaacs, "The Standardizing of Contracts," *Yale Law Journal* 27:34–48 (1917–1918); and Roscoe Pound, "The End of Law as Developed in Juristic Thought," esp. pp. 209 ff.

2. Ibid.; *cf.* Morris R. Cohen, pp. 568–571.

3. Ibid., pp. 218, 213, 212.

4. Ibid., pp. 219–221.

5. Ibid., p. 221.

6. Robert L. Hale, "Bargaining, Duress, and Economic Liberty," *Columbia Law Review* 43:603–628 (July 1943), esp. pp. 625–628.

7. Isaacs, "The Standardizing of Contracts," pp. 39–40.

8. Ibid., p. 40.

9. Isaiah Berlin, *Two Concepts of Freedom* (Oxford, 1958).

10. Isaacs, "The Standardizing of Contracts," p. 47.

11. It is a moot question whether the cyclical theory is likely to apply in the future. It is hard to imagine a reversion to "freedom of contract" even in the U.S. in the twenty-first century.

12. Kessler and Sharp, pp. 19–20.

13. Ibid., p. 20.

14. Ibid.

15. Ibid., pp. 56 ff.; for a related point see Friedrich Kessler, "Contracts of Adhesion—Some Thoughts about Freedom of Contract," *Columbia Law Review* 43:629–642 (July 1943), esp. p. 635.

16. Ibid., pp. 80–81.

17. *Corbin on Contracts* (1950), Section 19.

18. Kessler and Sharp, pp. 81–82.

19. W. Friedmann, "Changing Functions of Contract in the Common Law," *University of Toronto Law Journal* 9:22–23 (1951). Repeated references in this chapter to government supervision, control, and regulation of business are not intended to imply I believe such government action is always or even frequently in fact in the public interest.

20.   Radin, "Contract Obligation."

21.   Patterson, p. 736.

22.   W. Friedmann, pp. 29–30.

23.   Ibid., p. 23.

24.   Karl N. Llewellyn, review of *The Standardization of Commercial Contracts in English and Continental Law,* by O. Prausnitz, *Harvard Law Review* 52:703 (1939).

25.   Radin, pp. 579–581, 583. Italics added.

26.   Kessler, "Contracts of Adhesion," pp. 640–641. Italics added.

27.   Arthur S. Miller, "Government Contracts and Social Control: A Preliminary Inquiry," *Virginia Law Review* 41:32 (1955). Italics added.

28.   A similar vision of the position of the Chinese government vis-à-vis planned contracts is suggested in Chap. IV above.

29.   Miller, p. 54.

30.   Much of this section on government contracting is drawn from ibid.; John W. Whelan and Edwin C. Pearson, "Underlying Values in Government Contracts," *Journal of Public Law* 10:298–344 (fall 1961); J. W. Whelan and J. T. Phillips, "Government Contracts: Emphasis on Government," *Law and Contemporary Problems* 29:315–346 (spring 1964); and Richard E. Speidel, "What Should the Law Schools Do About Federal Government Contracts?" *Journal of Legal Education* 18:371–394 (1966).

31.   Whelan and Pearson, p. 311.

32.   Speidel, "What Should the Law Schools Do," pp. 382–383. Speidel briefly describes the bind the contractor is in at p. 383.

33.   Of course, the government remains liable, but to a lesser degree, for its breach; see Whelan and Pearson, pp. 315–321.

34.   Ibid., p. 334. Italics added.

35.   Speidel, "What Should the Law Schools Do," pp. 389, 390–391.

36.   Whelan and Pearson, pp. 324–326.

37.   Speidel, "What Should the Law Schools Do," p. 390. Italics added.

38.   Whelan and Pearson, pp. 339–343.

39. The term "contract system" is used in Carl F. Stover, "The Government Contract System as a Problem in Public Policy," *George Washington Law Review* 32:701–718 (April 1964).

40. Ibid., pp. 707, 708, 710. Original italics deleted.

41. Ibid., p. 711.

42. Ibid., pp. 712–713. Italics added.

43. 312 F. 2d 418 (Ct. C., 1963), affirmed on rehearing, 320 F 2d 345 (Ct. Cl., 1963), *cert. denied*, 375 U.S. 954 (1964), *motion for rehearing denied*, 377 U.S. 1010 (1964), *second cert. denied*, 86 [Sup. Ct. 50] (1965).

44. Whelan and Phillips, p. 322.

45. Ibid., p. 337.

46. Quoted in ibid., p. 342.

47. Ibid.

48. Friedrich Kessler and Richard H. Stern, "Competition, Contract, and Vertical Integration," *Yale Law Journal* 69:1 (November 1959). This section of Chap. V on vertical integration is based essentially on this article, on a two-part article by Stewart Macaulay, "Changing a Continuing Relationship Between a Large Corporation and Those Who Deal With It: Automobile Manufacturers, Their Dealers, and the Legal System," *Wisconsin Law Review* (summer 1965), pp. 483–575 and (fall 1965), pp. 740–858; and on Friedrich Kessler, "Automobile Dealer Franchises: Vertical Integration by Contract," *Yale Law Journal* 66:1135–90 (July 1957).

49. Kessler and Stern, p. 2.

50. Macaulay, "Changing a Continuing Relationship," p. 846.

51. Kessler, "Automobile Dealer," pp. 1149–50.

52. Macaulay, "Changing a Continuing Relationship," p. 848.

53. Among those studies are Macaulay, "Changing a Continuing Relationship"; Stewart Macaulay, "Non-Contractual Relations in Business: A Preliminary Study," *American Sociological Review* 28:55–67 (February 1963); Macaulay, "Private Legislation and the Duty to Read—Business Run by IBM Machine, the Law of Contracts and Credit Cards," *Vanderbilt Law Review* 19:1051–1121 (October 1966); Franklin M. Schultz, "The Firm Offer Puzzle: A Study of Business Practice in the Construction Industry," *University of Chicago Law Review* 19:237–285 (winter 1952); Lawrence M. Friedman, *Contract Law in America*; and for general discussion see Lawrence M.

Friedman and Stewart Macaulay, "Contract Law and Contract Teaching: Past, Present, and Future," *Wisconsin Law Review* (fall 1967), pp. 805–821; Richard E. Speidel, "Contract Law: Some Reflections Upon Commercial Context and the Judicial Process," *Wisconsin Law Review* (fall 1967), pp. 822–832; and Addison Mueller, "Contract Remedies: Business Fact and Legal Fantasy," *Wisconsin Law Review* (fall 1967), pp. 833–839.

54. Max Weber, *On Law in Economy and Society,* Max Rheinstein, ed. (Cambridge, Mass., 1954).

55. Llewellyn, "What Price Contract," p. 723.

56. Ibid., p. 724, and Mueller, p. 836.

57. Ibid., p. 725.

58. Friedman and Macaulay, "Contract Law," p. 815.

59. Macaulay, "Non-Contractual Relations."

60. Ibid., pp. 60–61. Italics added.

61. Ibid., p. 61. Italics added.

62. Ibid.

63. Without resorting to legal sanctions, many effective sanctions are available within the business community in which the exchange relationship is embedded. There are business norms opposed to welshing on a deal, personal relationships that evolve, goodwill, good reputation, and the importance to the parties of the continuing relationship itself.

64. Bronislaw Malinowski, *Crime and Custom in Savage Society,* paperback ed. (Totowa, N.J., 1966), esp. pp. 24–68. Durkheim also was aware that complex exchange relationships do not live by contracts alone. He suggested that in addition to legally defined pressures inherent in contractually structured exchange relations, there are moral obligations and professional obligations, which while not legally enforceable are nonetheless in a sense imperative. This moral or customary noncontractual way of relating units in exchange relations need not be limited to the realm of professional morality. More generally, informal patterns of behavior grow up over time, which to a substantial extent may mold exchange relations without reference to legal contractual forms, even if the latter are available.

65. Barry M. Richman, *A Firsthand Study of Industrial Management in Communist China* (Los Angeles, 1967); Barry M. Richman, "Capitalists and Managers in Communist China," *Harvard Business Review* 45:57–58 (January-February 1967); Franz Schurmann, *Ideology and Organization;* and Audrey Donnithorne, *China's Economic System* (New York, 1967).

VI. *The Meaning of Contracts: The Question Posed*

1.   Loeber, pp. 131, 133.

2.   Ibid., p. 169.

BIBLIOGRAPHY

"Actively and Steadily Extending the Combined Contract Is One of the Most Powerful Measures for Promoting the Socialist Transformation of Agriculture," *Jen-min jih-pao* (People's daily; June 28, 1954).

"The Advantages and Understanding of Convoking Combined Conferences for Supply, Production, and Marketing," *Hsin-chiang jih-pao* (Sinkiang daily; Aug. 27, 1955).

"Agriculture and Commerce Conclude 'Five Fix' Production and Marketing Contracts," *Jen-min jih-pao* (May 3, 1961).

Basic Problems in the Civil Law of the People's Republic of China, see *Chung-hua jen-min kung-ho-kuo min-fa chi-pen wen-t'i.*

Berlin, Isaiah. *Two Concepts of Freedom.* Oxford, England, Clarendon Press, 1958.

Berliner, Joseph S. "The Informal Organization of the Soviet Firm," *The Quarterly Journal of Economics* 66:342–365 (August 1952).

Berman, Harold J. "Commercial Contracts in Soviet Law," *California Law Review* 35:191–225 (June 1947).

——— *Justice in Russia.* Cambridge, Mass., Harvard University Press, 1950.

——— "The Current Movement for Law Reform in the Soviet Union," *The American Slavic and East European Review* 15:179ff (1956).

——— "Soviet Law Reform—Dateline Moscow 1957," *Yale Law Journal* 66:1191ff (1957).

——— "Soviet Law and Government," *Modern Law Review* 21:19ff (January 1958).

——— *The Nature and Functions of Law.* Brooklyn, N.Y., Foundation Press, 1958.

——— "Force Majeure and the Denial of an Export License under Soviet Law: A Comment on 'Jordan Investments, Ltd. v. Soiuznefteksport'," *Harvard Law Review* 73:1128ff (April 1960).

——— Review of *Settling Disputes in Soviet Society: The Formative Years of Legal Institutions* by John N. Hazard, *Yale Law Journal* 70:685ff (March 1961).

"Bringing the Production and Marketing Activity of the Commune Production Brigade into State Plans, Various Places Widely Sign Production and Marketing Contracts," *Jen-min jih-pao* (Apr. 13, 1959).

"The Cabinet Approves and Transmits the Report of the Ministry of Commerce Demanding Level-by-level Convoking of Materials-Exchange Conferences," *Hsin-hua pan-yueh-k'an* (New China semi-monthly), no. 7 (1959).

"The Cabinet Clearly Fixes the Scope of the Free Market . . . ," *Ta kung pao* (Impartial daily; Aug. 18, 1957).

"Carry Out More Properly the Investment-by-Contract System," *Jen-min jih-pao* (Mar. 12, 1960), tr. in *Survey of China Mainland Press,* no. 2220:3.

Chang Yu-nan. "Industrial Administration in Communist China," *Western Political Quarterly* 9:850–872 (December 1956).

Chao Kuo-chun. *Economic Planning and Organization in Mainland China.* Cambridge, Mass., East Asian Research Center, Harvard University, 1959.

Ch'eng Tzu-hua. "Struggle to Further Develop the Exchange of Material between City and Town, to Promote the Movement to Increase Agricultural Production, Taking Mutual Aid and Cooperation as its Core, and to Aid Industrialization of the Country," *Jen-min jih-pao* (Nov. 30, 1954).

Chi Tsung. "Preliminary Experience in the Contract System of Investment," *Ts'ai-cheng* (Finance), no. 9 (Sept. 5, 1958), tr. in *Extracts from China Mainland Magazines,* no. 147:68–71.

"The Ch'iao (?)-k'ou SMCs Spreading the Combined Contract Is Not a Coincidental Matter," *Hsin Hunan pao* (New Hunan daily; May 15, 1957).

"Chin-hua Wholesale Station, Concluding Order Contracts with Local Handicraft Producers' Cooperatives, Improved the Production of the Producers' Cooperatives and Fulfilled the Supply and Marketing Cooperatives' Requirements for a Source of Goods," *Ta kung pao* (Tientsin, Nov. 13, 1953).

China News Analysis, 1957–1962.

Chung-hua jen-min kung-ho-kuo min-fa chi-pen wen-t'i (Basic problems in the civil law of the People's Republic of China). Institute of Civil Law, Central Political-Legal Cadres' School. Peking, Fa-lu ch'u-pan-she, 1958. Tr. United States Joint Publications Research Service (JPRS): 4879, CSO: 1473–5 (1961).

Cohen, Jerome Alan. "The Criminal Process in the People's Republic of China: An Introduction," *Harvard Law Review* 79.3:469–533 (January 1966).

——— "Chinese Mediation on the Eve of Modernization," *California Law Review* 54:1201ff (August 1966).

Cohen, Morris R. "The Basis of Contract," *Harvard Law Review* 46:553–592 (February 1933).

"The Commercial Departments in Wan-hsien Special District Completely Carry Out the Contract System," *Ta kung pao* (Mar. 22, 1961).

"Concluding Speech Made by Chin Ming, Vice-Minister of Finance at National Conference for the Exchange of Experiences Regarding the System of Investment-by-Contract for Capital Construction," *Ts'ai-cheng,* no. 5 (March 1960), tr. in U.S. Joint Publication Research Service, no. 2984.

"Continue to Implement the Supply and Marketing Combined Contract System," *Nan-fang jih-pao* (Southern daily; Nov. 9, 1954).

Corbin on Contracts (1950), 1, Section 19.

"Decision of the Central Committee of the Chinese Communist Party and the Cabinet, Regarding Improving the System of Rural Finance and Trade Administration in Response to the Conditions of Communization," *Hsin-hua pan-yueh-k'an* (Nov. 1, 1959), p. 63.

"The Directive Issued by the Cabinet for the Work of the Advance Purchase of Cotton," *Jen-min jih-pao* (Dec. 23, 1955).

"The Directive Issued by the Cabinet for This Year's Advance Purchase of Cotton," *Jen-min jih-pao* (Mar. 23, 1955).

"The Directive of the Central Committee of the Chinese Communist Party and the Cabinet Relating to Organizing Rural Market Trade," *Ta kung pao* (Sept. 25, 1959).

"Directive of the Headquarters Branch of the People's Banks and the Grain Ministry Regarding Doing Well the Work of Supplying Capital for Summer Edible Oil Purchases," *Jen-min shou-ts'e* (People's handbook), p. 214. Peking, 1962.

Donnithorne, Audrey. *China's Economic System.* New York, Praeger, 1967.

Durkheim, Emile. *The Division of Labor in Society.* New York, Free Press, paperback, 1964.

Escarra, Jean. *Chinese Law.* Peking and Paris, 1936. Tr. G. R. Browne, for Works Progress Administration, W.P. 2799. Seattle, University of Washington, 1936. Xerox reprint, Cambridge, Mass., Harvard Law School and East Asian Research Center, 1961. Originally titled *Le Droit Chinois: Conception et Evolution, Institutions Legislatives et Judiciares, Science et Enseignement.*

"Extend the Combined Contract between the Mutual Aid Teams and the Cooperatives," *Jen-min jih-pao* editorial (Apr. 10, 1952).

"The First Whole Country Materials-Exchange Conference for Third Category Materials," *Jen-min jih-pao* (Mar. 23, 1959), as reprinted in *Hsin-hua pan-yueh-k'an,* no. 11:127 (1959).

Friedman, Lawrence M. *Contract Law in America.* Madison, University of Wisconsin Press, 1965.

——— and Stewart Macaulay. "Contract Law and Contract Teaching: Past, Present and Future," *Wisconsin Law Review* (fall 1967), pp. 805–821.

Friedmann, W. "Changing Functions of Contract in the Common Law," *University of Toronto Law Journal* 9:15–41 (1951).

Fuller, Lon L. *Basic Contract Law.* St. Paul, Minn., West Publishing Co., 1947.

——— "Pashukanis and Vyshinsky: A Study in the Development of Marxian Legal Theory," *Michigan Law Review* 47:1157–65 (June 1949).

——— and W. R. Perdue. "The Reliance Interest in Contract Damages," *Yale Law Review* 46:52–96 (November 1936); 46:373–420 (January 1937).

136

Green, Philip. "Science, Government, and the Case of RAND: A Singular Pluralism," *World Politics* 20:301-326 (January 1968).

Gsovski, Vladimir. *Soviet Civil Law.* Ann Arbor, University of Michigan Law School, 1948.

――― and Kazimierz Grzybowski. *Government, Law and Courts in the Soviet Union and Eastern Europe.* New York, Praeger, 1959.

Hale, Robert L. "Bargaining, Duress, and Economic Liberty," *Columbia Law Review* 43:603-628 (July 1943).

Hampsch, George H. 'Marxist Jurisprudence in the Soviet Union: A Preliminary Survey," *Notre Dame Lawyer* 35:525-535 (1960).

Havighurst, Harold C. *The Nature of Private Contract.* Evanston, Ill., Northwestern University Press, 1961.

"Holding Materials-Exchange Conferences Strengthens the Planned Nature of Production and Circulation," *Ta kung pao* editorial (Mar. 23, 1959).

Hooton, G. L. V. "The Planning Structure and the Five Year Plan in China," *Contemporary China,* vol. I. Hong Kong, Hong Kong University Press, 1955.

"How to Develop Contract Management," *Jen-min jih-pao* (Apr. 1, 1950).

"Hsi Liang Large Brigade Signs an Advance Purchase Contract for Cotton with the State," *Ta kung pao* (Feb. 2, 1963).

Hsia, Ronald. *Economic Planning in Communist China.* New York, International Secretariat, Institute of Pacific Relations, 1955.

Hsia Yun. "Investment Contract System, A Revolution in the Management of Capital Construction," *Hung-ch'i* (Red flag), no. 6 (Mar. 16, 1960), tr. in *Extracts from China Mainland Magazines,* no. 207:22-28.

Hsiao, Gene. "The Role of Economic Contracts in Communist China" *California Law Review* 53:1029-60 (October 1965).

Hsieh Wu. "Discuss the Contract System," *Cheng-fa yen-chiu* (Political-legal research), no. 2:43 (1959).

Hughes, T. J. and D. E. T. Luard. *The Economic Development of Communist China, 1949-1958.* London, Oxford University Press, 1959.

"In Purchasing Agricultural and Sideline Products . . . ," *Ta kung pao* (Feb. 15, 1962).

"Introduce the Connecting Contract between the An Yang Wu Kuan Supply and Marketing Cooperative, Credit Cooperative and Agricultural Producers' Cooperative," *Ta kung pao* (Tientsin, May 25, 1954).

"Investment in Capital Construction Must Be Placed on a Contract System," *Jen-min jih-pao* (July 15, 1958), tr. in *Survey of China Mainland Press,* no. 1822:2-4.

Isaacs, Nathan. "The Standardizing of Contracts," *Yale Law Journal* 27:34-48 (1917-1918).

――― "John Marshall on Contracts: A Study in Early American Juristic Theory," *Virginia Law Review* 7:413-428 (March 1921).

"It Is Necessary to Seasonably Arrange the Work of Advance Purchase of Cotton," *Jen-min jih-pao* (Dec. 23, 1955).

Jen Chien-hsin. "Strengthen the Economic Contract Work and Promote the Smooth Execution of National Economic Planning," *Cheng-fa yen-chiu*, no. 1:31 (1957).

Kessler, Friedrich. "Contracts of Adhesion—Some Thoughts about Freedom of Contract," *Columbia Law Review* 43:629–642 (July 1943).
——— "Automobile Dealer Franchises: Vertical Integration by Contract," *Yale Law Journal* 66:1135–90 (July 1957).
——— and Malcolm Sharp. *Contracts, Cases and Materials.* New York, Little, 1953.
——— and Richard H. Stern. "Competition, Contract, and Vertical Integration," *Yale Law Journal* 69:1ff (November 1959).
Kirby, E. S. "The Central Structure of Economic Planning in Communist China," *Far Eastern Economic Review* 22:737–741, 801–805 (June 13, 1957).
Kuan Ta-t'ung. "Discuss Rural Market Trade," *Jen-min jih-pao* (Nov. 21, 1959).
——— "My Country's Unified Socialist Domestic Market," *Hung-ch'i*, no. 6: 28–35 (1963).

Lewis, John Wilson. *Leadership in Communist China.* Ithaca, N.Y., Cornell University Press, 1963.
Li Che-jen. "The Combined Contracts of the SMCs, APCs, and Mutual Aid Teams of Two Villages," *Hsin-hua yueh-k'an* (New China monthly; May 1952), pp. 107–108.
Li Ch'eng-shui, et al. "Broadly Carry Out the Contract System," *Ta kung pao* (Feb. 20, 1959).
Li Choh-ming. "The First Decade, Part II, Economic Development," *The China Quarterly* 1:35–50 (January-March 1960).
——— *The Statistical System of Communist China.* Berkeley and Los Angeles, University of California Press, 1962.
Li Fu-ch'un on State Planning, New China News Agency (Sept. 24, 1956), tr. in *Survey of China Mainland Press,* no. 1378:3.
Lin I-chou. "Contract of Production Must Be Practical," *Hung-ch'i,* no. 10 (1959), tr. in *Extracts from China Mainland Magazines,* no. 172:7.
Liu Ch'i-shan. "The Combined Contract is a New, Independent Contract System," *Cheng-fa yen-chiu,* no. 3:37–40 (1957).
Liu Yung-hua (Lou Hua). "The Lock and Chain on the Neck of the Mainland Small Trader—Discussing the Chinese Communists' Cooperative Undertaking of the Supply and Marketing 'Combined Contract' System," *Tsu-kuo chou-k'an* (Motherland weekly), 12.2:16 (Oct. 10, 1955).

138

——— (Yu Heng). "Commenting on the Chinese Communists' Rural Trade Market," *Tsu-kuo chou-k'an* (Nov. 2, 1959), p. 21.

——— (Lou Hua). "Communist Chinese Domestic Trade in 1961," *Tsu-kuo chou-k'an* (Apr. 30, 1962), pp. 12–17.

——— (Lou Hua). "Chinese Communist Domestic Trade in 1963," *Tsu-kuo chou-k'an,* no. 2:14–18 (1964).

Llewellyn, Karl N. "What Price Contract?—An Essay in Perspective," *Yale Law Journal* 40:704–751 (1931).

——— Review of *The Standardization of Commercial Contracts in English and Continental Law* by O. Prausnitz, *Harvard Law Review* 52:703 (1939).

Loeber, Dietrich A. "Plan and Contract Performance in Soviet Law," University of Illinois *Law Forum* (spring 1964), pp. 128–179.

"The Lung Chang Commercial Bureau and the Hu Chia Commune Conclude Purchase and Marketing Contracts; the State Plan Is Bound to the Commune Plan," *Ta kung pao* (Nov. 2, 1958).

Macaulay, Stewart. "Non-Contractual Relations in Business: A Preliminary Study," *American Sociological Review* 28:55–67 (February 1963).

——— "Changing a Continuing Relationship Between a Large Corporation and Those Who Deal With It: Automobile Manufacturers, Their Dealers, and the Legal System," *Wisconsin Law Review* (summer 1965), pp. 483–575; (fall 1965), pp. 740–858.

——— "Private Legislation and the Duty to Read—Business Run by IBM Machines, the Law of Contracts and Credit Cards," *Vanderbilt Law Review* 19:1051–1121 (October 1966).

Maine, Henry Sumner. *Ancient Law.* Pollack's ed., 1930. Boston, Beacon Press paperback, 1963.

"The Making of Advance Deposits in Advance Purchase of Cotton Must Be Combined with the Supply of Materials," *Ta kung pao* (Apr. 24, 1963).

Malinowski, Bronislaw. *Crime and Custom in Savage Society.* Totowa, N.J., Littlefield Adams, 1966.

"The Method of Drawing up Annual Plans Undergoes an Important Reform," *Chi-hua ching-chi* (Planned economy), no. 9 (Sept. 9, 1958), tr. in *Extracts from China Mainland Magazines,* no. 152:33–38.

Mihaly, Zoltan M. "The Role of Civil Law Institutions in the Management of Communist Economies: The Hungarian Experience," *American Journal of Comparative Law* 8:310ff (1959).

Miller, Arthur S. "Government Contracts and Social Control: A Preliminary Inquiry," *Virginia Law Review* 41:27–58 (1955).

"Ministries of Metallurgical Industry and Finance Point Out at Conference that Contract Investment System is the Correct Direction for Capital Construction Management," New China News Agency (July 14, 1958), tr. in *Survey of China Mainland Press,* no. 1822:4–5.

Mueller, Addison. "Contract Remedies: Business Fact and Legal Fantasy," *Wisconsin Law Review* (fall 1967), pp. 833–839.

"The 1952 Plan and the Contracts Concluded between the Erh Tao K'ou SMC . . . and the Hsieh Yun Hu Mutual Aid Team," *Jen-min jih-pao* (Apr. 10, 1952).

"1954 Charter of the All-China Headquarters Supply and Marketing Cooperative in Peking," *Jen-min jih-pao* (Nov. 30, 1954).

"179 Production and Marketing Vegetable Combined Contracts were Concluded in the City: A Great Deed in the Work of Reforming the Production, Supply and Marketing of Vegetables," *Changkiang jih-pao* (Nov. 21, 1956).

"Our Cities Develop in Large Amounts the Combined Contract for Vegetables," *Lü-ta jen-min jih-pao* (Lü-ta people's daily; Apr. 4, 1955[?]).

Patterson, Edwin W. "Compulsory Contracts in the Crystal Ball," *Columbia Law Review* 43:731–749 (July 1943).

"Pay Attention to Contract Management," *Ta kung pao,* date uncertain in 1957, in Union Research Institute files, 1957–1958, 84, no. 32165.

Perkins, Dwight H. *Marketing Control and Planning in Communist China.* Cambridge, Mass., Harvard University Press, 1966.

Pfeffer, Richard M. Review of *Ideology and Organization in Communist China* by Franz Schurmann, *Pacific Affairs* 39:349–360 (fall and winter, 1966–1967).

——— "Crime and Punishment: China and the United States," *World Politics* 21:152–181 (October 1968).

Pound, Roscoe. "Liberty of Contract," *Yale Law Journal* 18:454–487 (1909).

——— "The End of Law As Developed in Juristic Thought," *Harvard Law Review* 30:201–225 (January 1917).

——— "The New Feudalism," *American Bar Association Journal* 16:553–558 (1930).

"Production Contracts Must Follow the Mass Line," *Nan-fang jih-pao* (May 14, 1959), tr. in *Survey of China Mainland Press,* no. 2075:17–20.

"Promote Investment-by-Contract System for Capital Construction in Shanghai," *Chieh-fang jih-pao* (Liberation daily; May 5, 1960), tr. in *Joint Publications Research Service,* no. 5485:35–37.

"Provisional Measures of the Finance and Economic Committee of the Central People's Government Affairs Council Regarding the Signing of Contracts Between Organs, State-Managed Enterprises, and Cooperatives," *Chung-hua jen-min kung-ho-kuo fa-ling hui-pien* (Collection of laws and decrees of the Chinese People's Republic), p. 532. Peking, 1953.

"The Purchase and Marketing Combined Contract is Good," *Ta kung pao* (Jan. 30, 1962).

"The Purchase Price Is Not Sufficiently Logical; Some APCs Do Not Perform Their Contracts," *Ta kung pao* (Jan. 4, 1957).

"Purchase Work Can't Be Simplified," *Nan-fang jih-pao* (Mar. 20, 1963).

Quigley, John B., Jr. "Recent Comment on the Contract of Delivery by Soviet Writers," spring 1966, unpublished paper at the Harvard Law School.

Radin, Max. "Contract Obligation and the Human Will," *Columbia Law Review* 43:575–585 (July 1943).

"Rapidly Promote Investment-by-Contract System for Capital Construction," *Chieh-fang jih-pao* (May 5, 1960), tr. in Joint Publications Research Service, no. 5495:38–39.

"Re-evaluate Combined Production and Marketing Contracts for Live Pigs," *Chekiang jih-pao* (July 15, 1955).

Restatement of the Law of Contract. St. Paul, Minn., American Law Institute Publishers, student ed., 1933.

Richman, Barry M. "Capitalists and Managers in Communist China," *Harvard Business Review* 45:57–78 (January-February 1967).

——— *A Firsthand Study of Industrial Management in Communist China.* Los Angeles, Graduate School of Business Administration, University of California, 1967.

Schultz, Franklin M. "The Firm Offer Puzzle: A Study of Business Practice in the Construction Industry," *University of Chicago Law Review* 19:237–285 (winter 1952).

Schurmann, Franz. "China's 'New Economic Policy'—Transition or Beginning," *The China Quarterly*, no. 17:65–91 (January-March 1964).

——— *Ideology and Organization in Communist China.* Berkeley, University of California Press, 1966.

Schwartz, Benjamin I. *Chinese Communism and the Rise of Mao.* Cambridge, Mass., Harvard University Press, 1958.

"Settle Combined Production and Marketing Contracts for Live Pigs," *Che-chiang jih-pao* (July 15, 1955).

Skinner, G. William. "Marketing and Social Structure in Rural China," *The Journal of Asian Studies* 24.1:3–43 (November 1964); 24.2:195–228 (February 1965); 24.3:363–399 (May 1965).

"SMCs and APCs Should Conclude Combined Contracts," *Che-chiang nung-ts'un kung-tso t'ung-hsun* (Chekiang agricultural work bulletin), no. 49 (Mar. 4, 1955), as reprinted in *Chung-kuo nung-ts'un ti she-hui-chu-i kao-ch'ao* (Socialist high tide in Chinese agriculture), middle volume, pp. 679–680. Peking, People's Publishing House, 1956.

"Speech by Wang Hsueh-ming, Vice-Minister of Finance, before the National Conference for Exchange of Experiences Regarding the System of Investment by Contract for Capital Construction on January 11, 1960," *Ts'ai-cheng*, no. 5 (Mar. 9, 1960), tr. in Joint Publications Research Service, no. 2984, microfilm No.: 1960.

Speidel, Richard E. "What Should the Law Schools Do About Federal Government Contracts?" *Journal of Legal Education* 18:371–394 (1966).

——— "Contract Law: Some Reflections Upon Commercial Context and the Judicial Process," *Wisconsin Law Review* (fall 1967), pp. 822–832.

"Spread the Contract System and Do the Work of Supplying Production Materials Well," *Ta kung pao* (Tientsin, May 13, 1956).

Stone, Julius. "The Myths of Planning and Laissez Faire: A Re-Orientation," *George Washington Law Review* 18:1–49 (December 1949).

Stover, Carl F. "The Government Contract System as a Problem in Public Policy," *George Washington Law Review* 32:701–718 (April 1964).

Strong, Anna Louise. "Some Comments on the Chinese People's Communes," *Peking Review* (June 12, 1964), pp. 17–21.

"The Supply and Marketing Cooperative of the Lu Ch'ing Country, Wei-tsu Town and Concerned Departments Closely Coordinated to Carry Out the Four-way Combined Contract to Help Agricultural Production . . . ," *Ta kung pao* (Feb. 25, 1962).

Szczepanik, E. F. "The Principles of the Economic Policy of Maoism," *Contemporary China*, vol. I. Hong Kong, Hong Kong University Press, 1955.

"T'ai Yuan City Supply and Marketing Cooperative, After Investigation and Finding out the Causes for Problems Existing in Executing Contracts, Fixed Improved Methods," *Ta kung pao* (Tientsin, June 10, 1953).

"Take the Whole Country as a Coordinated Chess Game," *Jen-min jih-pao* (Feb. 24, 1959), tr. in *Survey of China Mainland Press*, no. 1970:1–7.

Tang, Peter S. H. *Communist China Today*, vol. I. 2nd ed. Washington, D.C., Research Institute on the Sino-Soviet Bloc, 1958.

Tawney, R. H. *Land and Labor in China.* New York, Harcourt Brace, 1932.

"To Guide the Communes to Carry Out Planned Production," *Ta kung pao* (Jan. 19, 1959).

Tseng Shan. "Special Problems Regarding Processing and Orders," *Kung-shang-chieh yueh-k'an* (Industrial and commercial monthly; Mar. 10, 1955).

Ts'ui Chih-min. "Expand the Purchase and Marketing of Native Products, and Extend Contact Management," *Chin-pu jih-pao* (Advance daily; Feb. 24, 1951).

Ts'ui Li-ch'un. "Several Problems That Should Be Attended to and the Function of the Contract System with Regard to Rural Sideline Production and Marketing," *Chung-kuo nung-pao* (Chinese agricultural news), 1.6: 67 (1950).

Tu Hsiang-kuang. "The Business Accounting System, A Prerequisite to Socialist Industrialization," *Hsueh-hsi* (Study), no. 6 (June 1955), tr. in *Survey of China Mainland Press*, no. 1106:35–42.

von Mehren, Arthur T. *The Civil Law System.* Englewood Cliffs, N.J., Prentice Hall, 1957.

――― "Civil Law Analogues to Consideration: An Exercise in Comparative Analysis," *Harvard Law Review* 72:1009ff (April 1959).

Weber, Max. *On Law in Economy and Society*, ed. Max Rheinstein. Cambridge, Mass., Harvard University Press, 1954.

Wei Fu-k'ai, et al. "Several Problems Regarding Implementing the Contract System," *Ta kung pao* (June 5, 1959).

Whelan, John W. and Edwin C. Pearson. "Underlying Values in Government Contracts," *Journal of Public Law* 10:298–344 (fall 1961).

Whelan, J. W. and J. T. Phillips. "Government Contracts: Emphasis on Government," *Law and Contemporary Problems* 29:315–346 (spring 1964).

"Widely Connecting, Actively Selling, Signing Contracts to Guide Production . . . ," *Ta kung pao* (Jan. 11, 1964).

"Widely Spread and Realistically Perform Combined Contracts," *Yunnan jih-pao* (June 18, 1956).

Wu Yuan-li. *Economic Survey of Communist China.* New York, Bookman Associates, 1956.

――― "Communist China's Industrialization at the Crossroads," *Current History* (September 1961), pp. 151–156.

Yang, C. K. *The Chinese Family in the Communist Revolution.* Cambridge, Mass., M.I.T. Press, 1959.

Yang Ch'un-hsu and Li Cheng-jui. "Apply the Contract System Extensively," *Ta kung pao* (Feb. 20, 1959), tr. in *Survey of China Mainland Press*, no. 1980:8–29.

Yang-Ying-chieh. "On Unified Planning and Phased Controls of National Economy," *Chi-hua ching-chi*, no. 11 (Nov. 9, 1958), tr. in *Extracts from China Mainland Magazines*, no. 156:24–28.

Yen Wang-fu. "How to Realize the Purchase and Marketing Combined Contract," *Ta kung pao* (Mar. 30, 1962).

Zile, Z. L. "Remedies for Breach of Contractual Obligations in Soviet Law," student paper, Harvard Law School, Comparative Legal Research, 1960.

GLOSSARY

chi-shih 集市 (rural market)

ch'i-yeh 企业 (business enterprise)

chia-kung ho-t'ung 加工合同 (processing contract)

chieh-ho ho-t'ung 結合合同 (combined contract)

ching-chi ho-suan chih-tu 經济核算制度 (business accountability system)

ch'ing-li 清理 (to settle, to liquidate)

chü-t'i chih-hsing ho-t'ung 具体执行合同 (a concrete, executing contract)

ho-t'ung (chih-tu) 合同制度 (contract [system])

hsieh-i 协議 (agreement)

hsien 县 (county)

hu-chu-tsu 互助组 (mutual aid team [MAT])

i-kou ho-t'ung 議购合同 (negotiated purchase contract)

jen-min kung-she 人民公社 (people's commune)

143

kung-hsiao ho-tso-she 供銷合作社 (supply and marketing cooperative [SMC])

nung-yeh ho-tso-she 农业合作社 (agricultural producers' cooperative [APC])

p'ai-chia 牌价 (list price [state set])

p'ai-kou 牌购 (purchases at list price)

p'ei-ho 配合 (to coordinate)

shuo-fu 説服 (to persuade)

tui-hsien 兑現 (to realize, to make good)

t'ung-kou t'ung-hsiao 統购統銷 (unified purchase and sales)

tzu-yu shih-ch'ang 自由市場 (free market)

wu-tzu chiao-liu-hui 物資交流会 (materials exchange conference)

yü-kou ho-t'ung 預购合同 (advance purchase contract)

INDEX

145

HARVARD EAST ASIAN MONOGRAPHS

1. Liang Fang-chung, *The Single-Whip Method of Taxation in China*

2. Harold C. Hinton, *The Grain Tribute System of China, 1845–1911*

3. Ellsworth C. Carlson, *The Kaiping Mines, 1877–1912*

4. Chao Kuo-chün, *Agrarian Policies of Mainland China: A Documentary Study, 1949–1956*

5. Edgar Snow, *Random Notes on Red China, 1936–1945*

6. Edwin George Beal, Jr., *The Origin of Likin, 1835–1864*

7. Chao Kuo-chün, *Economic Planning and Organization in Mainland China: A Documentary Study, 1949–1957*

8. John K. Fairbank, *Ch'ing Documents: An Introductory Syllabus*

9. Helen Yin and Yi-chang Yin, *Economic Statistics of Mainland China, 1949–1957*

10. Wolfgang Franke, *The Reform and Abolition of the Traditional Chinese Examination System*

11. Albert Feuerwerker and S. Cheng, *Chinese Communist Studies of Modern Chinese History*

12. C. John Stanley, *Late Ch'ing Finance: Hu Kuang-yung as an Innovator*

13. S. M. Meng, *The Tsungli Yamen: Its Organization and Functions*

14. Ssu-yü Teng, *Historiography of the Taiping Rebellion*

15. Chun-Jo Liu, *Controversies in Modern Chinese Intellectual History: An Analytic Bibliography of Periodical Articles, Mainly of the May Fourth and Post-May Fourth Era*

16. Edward J. M. Rhoads, *The Chinese Red Army, 1927–1963: An Annotated Bibliography*

17. Andrew J. Nathan, *A History of the China International Famine Relief Commission*